opposing viewpoints®

SOURCES

nuclear arms

1988 annual

David L. Bender, *Publisher*
Bruno Leone, *Executive Editor*
M. Teresa O'Neill, *Senior Editor*
Bonnie Szumski, *Senior Editor*
Janelle Rohr, *Senior Editor*
Lynn Hall, *Editor*
Susan Bursell, *Editor*
Julie S. Bach, *Editor*
Thomas Modl, *Editor*
William Dudley, *Editor*
Robert C. Anderson, *Editor*

greenhaven press, inc.
San Diego, CA

© 1988 by Greenhaven Press, Inc.

ISBN 0-89908-535-0
ISSN 0748-2876

contents

Editor's Note
Opposing Viewpoints SOURCES provide a wealth of opinions on important issues of the day. The annual supplements focus on the topics that continue to generate debate. Readers will find that *Opposing Viewpoint SOURCES* become the barometers of today's controversies. This is achieved in three ways. First, by expanding previous chapter topics. Second, by adding new materials which are timeless in nature. And third, by adding recent topical issues not dealt with in previous volumes or annuals.

Viewpoints

"The ultimate goal should be a state of mutual deterrence at the lowest force levels consistent with stability."

US Nuclear Strategy Should Be Based on Deterrence

Robert S. McNamara

It is nearly fifty years since Albert Einstein sent his historical letter to President Roosevelt warning him that it was essential that the United States move quickly to develop the nuclear bomb. In that half-century the world's inventory of such weapons has increased from zero to fifty thousand. On average, each of them has a destructive power thirty times that of the Hiroshima bomb. A few hundred of the fifty thousand could destroy not only the United States, the Soviet Union, and their allies, but, through atmospheric effects, a major part of the rest of the world as well.

The weapons are widely deployed. They are supported by war-fighting strategies. Detailed war plans for their use are in the hands of the field commanders. And the troops of each side routinely undertake exercises specifically designed to prepare for that use. General Bernard Rogers, the Supreme Allied Commander of NATO forces in Europe, has said it is likely that in the early hours of a military conflict in Western Europe, he would in fact ask for the authority to initiate such use.

A Series of Decisions

This situation has evolved over the years through a series of incremental decisions. I myself participated in many of them. Each of the decisions, taken by itself, appeared rational or inescapable. But the fact is that they were made without reference to any overall master plan or long-term objective. They have led to nuclear arsenals and nuclear war plans that few of the participants either anticipated or would, in retrospect, wish to support.

Although four decades have passed without the use of nuclear weapons, and though it is clear that both the United States and the U.S.S.R. are aware of the dangers of nuclear war, it is equally true that for

Robert S. McNamara, in a speech delivered to the Economics Club of Detroit in Detroit, Michigan on February 17, 1987.

thousands of years the human race has engaged in war. There is no sign that it is about to change. And history is replete with examples of occasions in such wars when emotions have taken hold and replaced reason.

I do not believe the Soviet Union wants war with the West. And certainly the West will not attack the U.S.S.R. or its allies. But dangerous frictions between East and West have developed in the past and are likely to do so in the future. If deterrence fails and conflict develops, the present Western strategy carries with it a high risk that our civilization will be destroyed.

During the seven years I served as Secretary of Defense, confrontations carrying a serious risk of military conflict developed on three separate occasions: over Berlin in August of 1961; over the introduction of Soviet missiles into Cuba in October of 1962; and in the Middle East in June of 1967. In none of these cases did either side want war. In each of them we came perilously close to it.

It is correct to say that no well-informed, coolly rational political or military leader is likely to initiate the use of nuclear weapons. But political and military leaders, in moments of severe crisis, are likely to be neither well informed nor coolly rational.

Today we face a future in which for decades we must contemplate continuing confrontation between East and West. Any one of these confrontations can escalate, through miscalculation, into military conflict. And that conflict will be between blocs that possess fifty thousand nuclear warheads—warheads that are deployed on the battlefields and integrated into the war plans. A single nuclear-armed submarine of either side would unleash more firepower than man has shot against man throughout history.

In the tense atmosphere of a crisis, each side will feel pressure to delegate authority to fire nuclear weapons to battlefield commanders. As the likelihood of attack increases, these commanders will face a desperate dilemma: use the weapons or lose them. And because the strategic nuclear forces and the complex systems designed to command and control them, are perceived by many to be vulnerable to a preemptive attack, they will argue the advantage of a preemptive strike.

But it is a fact that in the face of the Soviet nuclear forces the West has not found it possible to develop plans for the use of its own nuclear weapons in a conflict with the U.S.S.R. in ways that would both assure a clear advantage to the West and at the same time avoid the very high risk of escalating to all-out nuclear war.

"A long-term, stable relationship between East and West is both desirable and attainable."

The risk that military conflict will quickly evolve into nuclear war, leading to certain destruction of our society, is far greater than I am willing to accept on military, political, or moral grounds. . . .

The conviction, therefore, that we must change course is shared by groups and individuals as diverse as the anti-nuclear movements, the majority of the world's top scientists, Soviet leader Mikhail Gorbachev and President Reagan, and such leaders of Third World and independent nations as Rajiv Gandhi and the late Olof Palme. All agree that we need a plan to reduce the long-term risk of nuclear war, but there is no consensus on what course to take. The changes of direction being advocated follow from very different diagnoses of our predicament.

Five quite different proposals have been presented to deal with the problem. They include:

—Achieving political reconciliation between East and West.

—Eliminating all nuclear weapons through negotiation (as proposed by General Secretary Gorbachev).

—Replacing "deterrence" with "defense"—the elimination of nuclear weapons by the substitution of defensive forces for offensive forces (as proposed by President Reagan).

—Strengthening deterrence by adding defensive forces to the offense (as proposed by Henry Kissinger and others).

—Accepting the proposition that nuclear warheads have no military use whatsoever except to deter one's opponent's use of such weapons.

Do any of these alternatives offer hope that the risk of nuclear war can be significantly reduced in the second half century of the nuclear age?

I will discuss each of them in turn beginning with East-West reconciliation.

The East-West military rivalry is, of course, a function of the political conflict that divides the two blocs. Many have argued, therefore, that any long-term attempt to bring a halt to the arms race and to reduce the risk of nuclear war must begin by addressing the source of the tensions—the political rivalry.

It is clear that the West—North America, Western Europe and Japan—lacks an agreed conceptual framework for the management of relations with the Soviet Union and its allies.

We need a coherent, widely supported policy, rooted in reality and pressed with conviction and determination. It must be a policy which protects our vital interests, enhances political cohesion, and offers the hope of influencing the Soviets to move in a favorable direction. A long-term, stable relationship between East and West is both desirable and attainable. Even in an atmosphere of competition and mutual suspicion there are common interests, and the pursuit of each side's competitive goals can take place in an atmosphere of moderation. . . .

Steps to control directly and reverse the arms race must go forward in parallel with efforts to reduce political tension.

A Nuclear-Free World

I turn, therefore, to consideration of the four different approaches to controlling directly the "volume" and "use" of such weaponry.

Mikhail Gorbachev, General Secretary of the Soviet Communist Party, has proposed that the United States and the Soviet Union aim at achieving the total elimination of nuclear weapons by the year 2000.

Is a nuclear-free world desirable if attainable? I believe it is, and I think most Americans would agree.

However, NATO's current military strategy and war plans are based on the opposite premise. And many—I would say most—U.S. military and civilian officials, as well as European leaders, hold the view that nuclear weapons are a necessary deterrent to Soviet aggression with conventional forces. Thus, these individuals do not favor a world without nuclear weapons. Zbigniew Brzezinski, President Carter's national security advisor, said of Gorbachev's proposal, "It is a plan for making the world safe for conventional warfare. I am therefore not enthusiastic about it."

My criticism of Gorbachev's vision, however, is not that it is undesirable, but that it is infeasible under foreseeable circumstances.

Unless we can develop technologies and procedures to ensure detection of any steps toward building a single nuclear bomb by any nation or terrorist group, an agreement for total nuclear disarmament will almost certainly degenerate into an unstable rearmament race. Thus, despite the desirability of a world without nuclear weapons, an agreement to that end does not appear feasible either today or for the foreseeable future.

SDI

On March 23, 1983, President Reagan proposed his solution to the problem of security in the nuclear age. He launched the Strategic Defense Initiative (SDI), a vast program that promised to create an impenetrable shield to protect the entire nation against a missile attack. With the shield in place, the President argued, we would be able to discard not just nuclear deterrence but nuclear weapons themselves.

The President continues to promise that this strategic revolution is at hand.

Virtually all others associated with the SDI have recognized and admitted that such a leakproof defense is so far in the future, if indeed it ever proves feasible, that it offers no solution whatsoever to our present dilemma. Therefore, they are advocating missions for a Star Wars system other than a perfect "security shield." These alternative aims range from defense of hardened targets—for example, missile silos and command centers—to partial protection of our populations.

For the sake of clarity I will call these alternative programs Star Wars II, to distinguish them from the President's original proposal, which will be labeled Star Wars I. It is essential to understand that Star Wars I and Star Wars II have diametrically opposite objectives. The President's program, if achieved, would substitute defensive for offensive forces. In contrast, Star Wars II systems have one characteristic in common: they would all require that we continue to maintain offensive forces but add the defensive systems to them. . . .

"It would be foolhardy to dismiss as mere propaganda the Soviets' repeated warnings that a nationwide U.S. strategic defense is highly provocative."

The most powerful argument put forward by those who favor "offense plus defense" is that presented by Kissinger: even a partially effective defense would introduce an element of uncertainty into Soviet attack plans and would thereby enhance deterrence. This assumes that the Soviet military's sole concern is to attack us and that any uncertainty in their minds is therefore to our advantage. But any suspicions they may harbor about our wishing to achieve a first-strike capability—and they do indeed

hold such views—would be inflamed by a partially effective defense.

Why will the Soviets suspect that Star Wars II is designed to support a first-strike strategy? Because a leaky umbrella offers no protection in a downpour but is quite useful in a drizzle. That is, such a defense would collapse under a full-scale Soviet first strike but might cope adequately with the depleted Soviet forces that had survived a U.S. first strike.

The Soviet Interpretation

And that is what causes the problem. President Reagan, in a little-remembered sentence in his March 23, 1983, speech, said, "If paired with offensive systems, (defensive systems) can be viewed as fostering an aggressive policy, and no one wants that." The President was concerned that the Soviets would regard a decision to supplement—rather than replace—our offensive forces with defenses as an attempt to achieve a first-strike capability. Reagan has subsequently said, "I think that would be the most dangerous thing in the world, for either one of us to be seen as having the capacity for a first strike." But that is exactly how the Soviets are interpreting our program. . . .

It would be foolhardy to dismiss as mere propaganda the Soviets' repeated warnings that a nationwide U.S. strategic defense is highly provocative. Their promise to respond with a large offensive buildup is no empty threat. Each superpower's highest priority has been a nuclear arsenal that can assuredly penetrate to its opponent's vital assets. Such a capability, each side believes, is needed to deter the other side from launching a nuclear attack or using a nuclear advantage for political gain.

We have said we would respond to a Soviet strategic defense plan in exactly the same way they have stated they would respond to ours.

We can safely conclude, therefore, from both the U.S. and Soviet statements, that any attempt to strengthen deterrence by adding strategic defenses to strategic offensive forces will lead to rapid escalation of the arms race. . . .

We are left, then, to turn to our final option: a reexamination of the military role of nuclear weapons.

Earlier I stated that no one had ever developed a plan for initiating the use of such weapons with benefit to the West. . . .

If there is a case for NATO retaining its present strategy, that case must rest on the strategy's contribution to the deterrence of Soviet conventional force aggression being worth the risk of nuclear war in the event deterrence fails.

But as more and more Western political and military leaders recognize, and as they publicly avow, the launch of strategic nuclear weapons against the Soviets' homeland—or even the use of battlefield nuclear weapons—would bring greater destruction to the West than any conceivable contribution they might make to its defense, there is less and less likelihood that the West would authorize the use of any nuclear weapons except in response to a Soviet nuclear attack. As this diminishing prospect becomes more and more widely perceived—and it will—whatever deterrent value still resides in the West's nuclear strategy will diminish still further. One cannot build a credible deterrent on an incredible action.

There are additional factors to be considered. Whether it contributes to deterrence or not, the threat of first use is not without its costs. It is a most contentious policy, leading to divisive debates both within Western Europe and North America; it reduces the West's preparedness for conventional war; and, as I have indicated, it greatly increases the risk of nuclear war.

"Our present nuclear policy is indeed bankrupt."

The costs of whatever deterrent value remains in the West's nuclear strategy are substantial.

Now, couldn't equivalent deterrence be achieved at lesser "cost"? I believe the answer is yes. Compared to the huge risks which we now run by relying on increasingly less credible nuclear threats, recent studies have pointed to ways by which the conventional forces may be strengthened at modest military, political, and economic cost.

The West has not done so because there is today no consensus among its military and civilian leaders on the military role of nuclear weapons.

There is, however, a slow but discernible movement toward acceptance of three facts:

—The West's existing plans for initiating the use of nuclear weapons if implemented are far more likely to destroy Western Europe, North America, and Japan than to defend them.

—Whatever deterrent value remains in the West's nuclear strategy is eroding and is purchased at heavy cost.

—The strength, and hence the deterrent capability, of Western conventional forces can be increased substantially within realistic political and financial constraints.

Mutual Deterrence

It is on the basis of these facts that I propose that we accept that nuclear warheads have no military use except to deter one's opponent from their use— and I suggest further that, for the long run, we base all our military plans, our defense budgets, our weapons development and deployment programs,

and our arms negotiations on that proposition.

The ultimate goal should be a state of mutual deterrence at the lowest force levels consistent with stability.

If the Soviet Union and the United States were to agree, in principle, that each side's nuclear force would be no larger than was needed to deter a nuclear attack by the other, how might the size and composition of such a limited force be determined?

When discussing Gorbachev's proposal for the total elimination of nuclear weapons, I pointed out that a nuclear-free world, while desirable in principle, was infeasible under foreseeable circumstances because the fear of cheating in such an agreement would be very great indeed. I stressed, however, that policing an arms agreement that restricted each side to a small number of warheads is quite feasible with present verification technology. The number required for a force sufficiently large to deter cheating would be determined by the number the Soviets could build without detection by NATO. I know of no studies which point to what that number might be, but surely it would not exceed a few hundred, say at most five hundred. . . .

We can—indeed we must—move away from the ad hoc decision making of the past several decades. It is that process which has led to a world in which the two great power blocs, not yet able to avoid continuing political conflict and potential military confrontation, face each other with nuclear war-fighting strategies and nuclear arsenals capable of destroying civilization several times over.

Most Americans are simply unaware that Western strategy calls for early initiation of the use of nuclear weapons in a conflict with the Soviets. Eighty percent believe we would not use such weapons unless the Soviets used them first. They would be shocked to learn they are mistaken. And they would be horrified to be told that senior military commanders themselves believe that to carry out our present strategy would lead to destruction of our society.

But those are the facts.

In truth, the Emperor has no clothes. Our present nuclear policy is indeed bankrupt.

President Reagan's intuitive reaction that we must change course—that we must recognize nuclear warheads cannot be used as military weapons—is correct. To continue as in the past would be totally irresponsible.

Robert S. McNamara was the US secretary of defense during the Kennedy and Johnson administrations in the 1960s. He then served as president of the World Bank until 1981.

"Effective strategic defense would pave the way for diminishing . . . the danger posed by nuclear weapons."

US Nuclear Strategy Should Be Based on Defense

Caspar W. Weinberger

Today, the United States faces many challenges as the leader of the free world. But one of the most pressing challenges must be confronted on the intellectual level. It involves America's willingness to question its assumptions in light of the practical results they have had in the world. America needs to ask if conventional wisdom is always wise.

In the realm of U.S. defense strategy, it is necessary to move away from partisan politics and toward a serious evaluation of our strategy's basic tenets, especially since the advent of the idea of mutual assured destruction in the 1960s. For nearly 20 years the United States has narrowed its options significantly by always saying no to the possession of defensive systems.

It is time that America examines the results of that decision. Obviously, elements of that overall strategy, such as the basic U.S. commitment to nuclear deterrence and the NATO strategy of flexible response, retain proven value. But by examining its strategic principles in an intellectually honest manner, the United States can reaffirm those tenets that still apply and revise any that have become outdated. The notion that abandonment of defenses has been stabilizing is one that deserves particular attention.

History's Parallel

History provides a fascinating parallel to the current controversy over whether defense should be considered part of U.S. nuclear strategy. In 1934 a British parliamentary debate, not unlike the present debate over the Strategic Defense Initiative (SDI), took place on the need for research in air defense technology. At that time, there was a growing concern about Nazi Germany's efforts to rearm itself. Many people believed that the potential

Caspar W. Weinberger, "Why Offense Needs Defense." Reprinted with permission from FOREIGN POLICY 68 (Fall 1987). Copyright 1987 by the Carnegie Endowment for International Peace.

destructiveness of bombing civilian populations from airplanes had been demonstrated clearly in World War I and that no defense ever could be built to meet such an attack in the future. One proposed solution was to build a bomber force that would act as a deterrent by jeopardizing Germany's own population.

However, Winston Churchill, then a member of Parliament and later prime minister, spoke in favor of air defense, saying, "If anything can be discovered that will put the earth on better terms against this novel form of attack, this lamentable and hateful form of attack—attack by spreading terror throughout civil populations—anything that can give us relief or aid in this matter will be a blessing to us all." He went on to argue that until defenses were devised, deterrence would rest completely on Great Britain's ability to retaliate: "The fact remains that pending some new discovery, the only direct measure of defense upon a great scale is the certainty of being able to inflict simultaneously upon the enemy as great damage as he can inflict upon ourselves."

Churchill's remarks reflected a sense both of realism and of hope. He was acutely aware of the technological limitations of his time. Yet he recognized the merit in seeking a deterrent that did not rely solely on offensive retaliation.

Since Churchill's time, technology has yielded the most destructive weapons ever known. Indeed, the advent of nuclear weapons has revolutionized the way people think about war, for its consequence is, quite possibly, the destruction of the world. . . .

A Different Strategy

America faces an adversary with a different strategic outlook from its own—an adversary that, while understanding the destructive potential of nuclear weapons, still plans to fight and win should either a nuclear or a conventional war break out.

This difference in strategic approach was evident in Soviet Marshal Vasily Sokolovsky's 1963 book *Military Strategy*. He and his collaborators made it clear that a nuclear war would most likely have to be fought. Indeed, because of such a possibility, planning for all contingencies was accomplished in a very calculated way. In light of technological developments since the nuclear revolution in military affairs, high-ranking Soviet officers like Marshal Nikolai Ogarkov, chief of the Soviet General Staff, and Colonel General M.A. Gareyev, chief of the military science directorate of the General Staff, have revised some of Sokolovsky's emphasis on early use of nuclear weapons. They now believe that it is possible to defeat the West through a nonnuclear offensive but have by no means ceased planning to prevail on the conventional or nuclear level "should imperialism unleash" a new war.

"One of the U.S. assumptions not shared by the Soviet Union, and whose profound effects are still being felt today, involved the idea of deterrence."

Although many contend that strategy does not apply to nuclear warfare in the same way it does to conventional warfare, recent history suggests that how one thinks about nuclear war decisively affects force deployments, force structure, weapons acquisition, and, above all, arms reduction agreements. The United States and the Soviet Union indeed have avoided war on the grand scale, even though preserving the peace has required a constant effort by the West to maintain its deterrent force and resolve. But no one can say with absolute certainty why these efforts have been successful thus far. Whatever the reason, America must continually search for better, more stable ways to keep the peace. This means thinking strategically. . . .

As the 1950s came to a close, the United States was still embroiled in a debate over how to avoid nuclear war. If the legacy of the 1950s was the emergence of a new approach toward strategy, the legacy of the 1960s might be described as the U.S. attempt to educate the Soviet Union to its way of thinking about the unthinkable. Indeed, a host of new ideas and concepts were carefully incorporated into the American nuclear lexicon. Consider the list: flexible response, escalation control, assured destruction, strategic stability, and escalation ladder, to name a few.

In a world of American nuclear pre-eminence, Soviet students of the United States were obliged at least to listen, even if somewhat incredulously. As is now known from Soviet military literature of the period, the Soviet decision makers paid scrupulous attention to American pronouncements, but they continued to operate under a different set of assumptions. One of the U.S. assumptions not shared by the Soviet Union, and whose profound effects are still being felt today, involved the idea of deterrence based solely on offensive retaliatory capabilities. Then Secretary of Defense Robert McNamara firmly believed that if both the United States and the Soviet Union possessed a secure second-strike capability without defenses, this represented the most stable situation that could be hoped for in a nuclear age.

The Soviet View of Deterrence

During the now famous encounter with then Soviet President Aleksei Kosygin at Glassboro, New Jersey, in June 1967, McNamara expressed frustration at not being able to convince Kosygin to take a similar view to America's and abandon the quest for a defense against nuclear missiles. Kosygin reportedly remarked: "We are defending Mother Russia—that's moral. You are increasing your offensive forces—that's immoral." Apparently, one of the most basic and ancient principles of strategy— the need for a mix of offensive and defensive capabilities to deter potential enemies—had been discarded in American strategic thinking then and replaced by concepts preached by the new American nuclear strategists.

Certainly, part of the logic that guided American policymakers reflected the problems associated with the development of technologies for defense against ballistic missiles. Given the lead that the United States possessed in nuclear weaponry, the inclination toward a purely offensive deterrent strategy can be understood. In hindsight, however, the consequences of abandoning the more traditional approach to strategy can be easily discerned.

On the one hand, the Soviet Union proceeded to explore and develop defensive technologies; on the other, the United States allowed what offensive advantage it possessed in nuclear technology to erode while making no attempt to acquire defensive technologies. It made little effort even to conduct research. McNamara, in full good faith, intentionally allowed this to happen. As early as January 1967, in testimony before a Senate committee, he justified his reasoning for not pursuing strategic defense:

> We believe the Soviet Union has essentially the same requirement for a deterrent or "Assured Destruction" force as the U.S. Therefore, deployment by the U.S. of an ABM [antiballistic missile] defense which would degrade the destruction capability of the Soviets' offensive force to an unacceptable level would lead to expansion of that force. This would leave us no better off than we were before.

Unfortunately, the Soviets approached their national security from an entirely different point of view, so McNamara found Soviet spending on "sievelike" defensive systems nearly incomprehensible. He called it "fanatical" and said it could be explained by "their strong emotional reaction to the need to defend Mother Russia." A better explanation is that Soviet deterrence doctrine has never conformed to a purely offensive approach. The United States made the familiar mistake of assuming that the Soviets were just like Americans and thought the same way.

A Heretical Idea

In 1983, when President Ronald Reagan first announced his initiative to develop strategic missile defenses, the idea that active defense might play a constructive role in the strategic equation, and that it might lead to a more stable world, was greeted by many so-called defense experts as nothing less than heresy. Indeed, "strategic defense" was not part of the nuclear lexicon and therefore was immediately dismissed by some in Congress and others in the defense community at large. Today, however, the basic principle is rarely challenged except by those who believe that the conventional wisdom will never change. The debate in Congress is not whether the United States should work on strategic defense technologies, but how vigorously. . . .

In the last 10 years alone, the Soviet Union has spent 15 times as much on strategic defense as the United States. This effort has given the Kremlin today the advantage of the world's only deployed operational ABM system, the only operational antisatellite system, the most comprehensive, in-depth, and capable air defense system ever deployed, and an organization for passive defense of its leadership, population, and industrial assets—including hardened shelters for 175,000 Communist party and government leaders. In fact, these actions, along with extensive advanced research into new technologies for ballistic missile defense, belie current Soviet rhetoric. . . .

An Invulnerable Soviet Population

The European experience before World War II bears a remarkable resemblance to the present, especially in some of the attitudes that ascribe solely peaceful and defensive intentions to the Soviet Union's military build-up. If it is not careful, the democratic West might one day find itself in the same dangerously weak position that the Allies were in when Germany attacked in 1939. But it might not have the time to recover as it did in World War II because the West would be facing a strongly defended, nuclear-armed Soviet Union with vast conventional and strategic strength.

It is hard to envision any circumstances more threatening and dangerous to the free world than those in which its populations and military forces remained completely vulnerable to Soviet nuclear missiles while Soviet populations and military assets were substantially protected against retaliatory

forces. This is exactly the kind of situation that could result if America formulated its strategy in a vacuum, ignoring all the hard-to-explain actions of its most likely adversary, or ascribed to that adversary intentions that fit into America's approach, but not necessarily the USSR's.

Moreover, any success the United States has had at bringing the Soviets to the negotiating table has been based on regained U.S. military strength and the will to continue to deter the use of Soviet military might, including American efforts in the area of strategic defense. It is only strength, along with a realistic military strategy, that impresses the Soviet Union.

It is time to place the strategic debate in the proper context. The issue for the foreseeable future is not how to remain vulnerable to nuclear attack but rather, how to balance U.S. deterrent strategy by shifting more and more to a mix of defensive and offensive arms.

Reagan's Vision

On March 23, 1983, Reagan challenged the scientific community to develop the technology that would allow the United States to move away from a policy of deterrence based solely upon nuclear retaliation and toward one based also upon the capability actually to defend the United States and its allies:

> What if free people could live secure in the knowledge that their security did not rest upon the threat of instant U.S. retaliation to deter a Soviet attack, that we could intercept and destroy strategic ballistic missiles before they reached our own soil or that of our allies?
> I know this is a formidable technical task. . . . But isn't it worth every investment necessary to free the world from the threat of nuclear war?

Like Churchill in the 1930s, Reagan challenged the so-called conventional wisdom of his time and was met with a great deal of skepticism and even ridicule. . . .

"The Soviet Union has spent 15 times as much on strategic defense as the United States."

Today, America also must look ahead and muster the courage and commitment that will be required to secure the president's vision of a more stable world—one based on mutual survival, not mutual destruction. This is an achievable goal, and the administration's highest priority.

I believe not only that strategic defense is possible, but that it represents the greatest hope for the future. Moreover, I view the SDI as the means for creating the conditions that will lead to progressive, deeper arms reductions. The SDI, on the one hand, and significant reductions in offensive nuclear arms, on the other, represent two complementary approaches to a more stable world.

Naturally, many claim that efforts by one side or the other to expand and improve defenses will be accompanied by the development of more offensive weapons designed to defeat the new defenses. This would be true if the new defenses deployed could be countered easily by the proliferation of offensive weapons. However, an effective defense, such as the one envisioned by the president, could not be countered in this way. Consequently, such a defense would not encourage proliferation of offensive missiles. . . .

Reducing the Threat

Effective defenses would remove incentives for ballistic missile proliferation. When the president directed work to begin on the SDI, that work was not undertaken to threaten the Soviet Union or to gain military superiority. Rather, the SDI was designed to introduce greater safety and security into the U.S.-Soviet relationship. Effective strategic defense would pave the way for diminishing, and thereby reducing, the danger posed by nuclear weapons to all countries.

"Defensive systems combined with equitable and verifiable reductions in conventional forces would permit countries to rid themselves of nuclear weapons."

Further, there can be an orderly transition to a defense-oriented world combining deployment of defensive weapons with a compensating reduction of offensive weapons. If both sides agreed to destroy a certain percentage of their ballistic missiles as they deployed defensive weapons systems, they would be sending each other signals that neither side was seeking a strategic advantage. In the long run, defensive systems combined with equitable and verifiable reductions in conventional forces would permit countries to rid themselves of nuclear weapons. Such a new order would have to be reached over a transition period during which deterrence would continue to rest mainly on the threat of retaliation, simply because there is no other means now to keep the peace. . . .

The perennial question free peoples ask with regard to defense is, How much is enough? To this there can be no precise answer. A country's security is a function of the degree of risk a country is willing to accept. The less risk America accepts, the more secure it will be. A country can never be totally secure, but it can run risks that invite aggression. Increased security that reduces risk requires increased costs, and increased costs require popular support and a determined commitment to pay freedom's price.

The Dangers of Naiveté

In the 1930s, in the face of German rearmament, some European countries decided that unilateral restraint and appeasement were enough to keep them safe. The United States also failed to appreciate the need for defense preparedness to avoid war. These misjudgments proved catastrophic, and the world still bears the scars of that collective naiveté. Better allied preparedness might well have deterred the Nazis. Instead, more than 40 million lives were lost because the West was not strong enough. . . .

The West has a long tradition of unwillingness to face harsh realities because they are so repugnant and foreign to its civilized heritage. The lack of preparedness before two world wars amply illustrates the democratic tendency to play down, and even ignore, potential threats to freedom. But the Western democracies must never forget how long it takes to regain strength once lost, and how vital it is to keep strong. Small investments each year can relieve them of the need to gamble that they will have enough time to regain strengths already lost. Moreover, by ensuring that their investment of resources is guided by an effective strategy, they make the most of their limited resources. The technological prowess of the West, combined with the innovative application of resources that is possible in free societies, will ensure its survival.

Caspar W. Weinberger was US secretary of defense from 1981 to 1987.

"The Soviet Union is more serious about . . . engaging the United States in productive negotiations on arms control than ever before in history."

Soviet Arms Control Efforts Are Sincere

Marshall D. Shulman

In his book, *The Cycles of American History*, Arthur Schlesinger recalls a lecture by the British historian Sir Herbert Butterfield at Notre Dame University during the early Cold War years. Entitled "The Tragic Element in Modern International Conflict," Butterfield's talk suggested that the historiography of international conflict has characteristically gone through two phases.

In the first or "heroic" phase, historians portray a struggle of right against wrong, of good people resisting bad. Then, as time passes and emotions subside, historians enter the second, "academic" phase, when they seek to understand the motives of the other side, and to define the structural dilemmas that so often underlie great conflicts between masses of human beings. Thus he notes, the "higher historiography" moves on from melodrama to tragedy. Butterfield's two phases, prescient at the time he wrote, are useful for a retrospective examination of the course of Soviet-American relations. . . .

Hostile Camps

Recall that it was 40 years ago that the United States went through one of the most remarkable transformations in American politics. It was a period when the matrix of the Cold War was established—a period of heroic accomplishments and of serious mistakes.

Within the space of a few months there was a massive turnaround in U.S. policy, from a period of collaboration with the Soviet Union as the "gallant ally" that had contributed heroically and with great loss of life to the defeat of the Nazi armies, to an alarmed and belated response to the problems of the postwar world—particularly the emerging Soviet dominance in Eastern Europe and a perceived Soviet threat to the Balkans and to Western Europe. . . .

This confrontational view has persisted because relations between the United States and the Soviet Union are powerfully influenced by the interaction of complex political forces within each of these countries. On the U.S. side, three factors need particular emphasis in explaining why American policy toward the Soviet Union has tended to be less than rational:

Familiar psychological mechanisms, most often operating below the conscious level, tend to make very real conflicts of interest between the United States and the Soviet Union appear absolute and therefore intractable. The subject is shot through with emotions, areas of dark uncertainty, fear, prejudices, and primitive stereotypes. Much of what Americans would wish to know about the Soviet Union is not known and may be unknowable, and therefore preconceptions are projected into these dark areas—partly from hopes and partly from fears.

Some of the psychological concepts developed to deal with interpersonal relations suggest applications to the way Americans tend to perceive the Soviet Union. Anxiety originating in many sources, including the hazard of nuclear destruction; displacement; denial; and projection are obviously relevant factors. The familiar "we-they" phenomenon—the tendency to apply separate standards to the good "in group" and the bad "out group"—leads Americans to look with indulgence on their own actions and with harsh severity on the Soviets': Soviet military programs reflect hostile intentions; American military programs are defensive. The expansion of Tsarist Russia and the Soviet Union to the Pacific proves Russian and Soviet intentions to conquer the world; the continental expansion of the United States was a matter of right. Soviet activities in the Third World

Marshall D. Shulman, "Four Decades of Irrationality: US-Soviet Relations," *Bulletin of the Atomic Scientists*, November 1987. Reprinted by permission of the BULLETIN OF THE ATOMIC SCIENTISTS, a magazine of science and world affairs. Copyright © 1987 by the Educational Foundation for Nuclear Science, 6042 S. Kimbark Ave., Chicago, IL 60637.

are manifestations of aggression; U.S. interventions are altruistic. Soviet espionage is traitorous; American espionage is patriotic. (This double standard, of course, applies with equal force to Soviet perceptions.) An obvious consequence of this "we-they" mechanism is that public discussion on this subject in the United States is dominated by simplified and polarized stereotypes.

It may be inevitable that people and leaderships should regard other nations in this way, but a greater awareness of the operation of psychological mechanisms can free us to distinguish what is real from what is fancied in our perception of each other, and to move in some small measure toward greater objectivity. It is a fundamental condition for the transition from Butterfield's first phase to the second, more analytical phase.

U.S. policy toward the Soviet Union has been greatly influenced by the vicissitudes of American politics. Since the mid-1970s the United States has experienced a conservative political tide in domestic politics and a resurgence of nationalism in foreign policy. The external ideological focus of this mood has been expressed in anticommunism as the main organizing principle for foreign policy; it has created a reaction against what is viewed as the weakness of the "liberal illusions" of the previous period by bringing about support for policies of greater activism and military strength against a perceived heightened threat from the Soviet Union. A hardening of popular attitudes toward the Soviet Union resulted from that nation's moves in Angola, Ethiopia, and Afghanistan, as well as from Moscow's actions against human rights and from continuing military buildup. These feelings were intensified by a backlash from the unrealistic expectations aroused by the détente period, as well as by a widespread sense of impotence stemming from the United States' experience in Vietnam and with the hostages in Iran, and by the passing of the period of American military superiority.

"The mindless increase in weapons programs, driven by parochial interests on both sides, has locked the two countries into a rising spiral."

The intellectual foundation of these sentiments derived from the neoconservative movement, which returned to the ideological fundamentalism of the postwar period in its approach to the Soviet Union, but is generally nationalist rather than internationalist in its foreign policy. In the prevailing climate, these views created powerful political support for a strong anti-Soviet posture, and they

discredited policies directed toward arms control, reduced tension, and [added] measures of cooperation with the Soviet Union. The most serious liability a politician could incur was to be labelled soft on communism and weak on defense.

The third factor to determine U.S. policy toward the Soviet Union has been the absence of rationality in decision making on military policy, the autonomy of the military establishment, and its increasing influence on U.S. foreign policy and domestic society. Decisions on defense policy, from research and development to the acquisition of weapons systems and deployment decisions, result from the interplay of parochial pressures and interests, rather than from an overarching determination of the national interest. Even if a president were to seek disinterested but competent counsel on defense policy concerns, there has been no mechanism to provide for such counsel since the abolition of the President's Science Advisory Committee. Today, such policy decisions depend on the outcome of bargaining between the military services, on the influence of the defense contractors, and on the economic and political interests of congressional districts.

Inflated Estimates of Soviet Power

Defense expenditures, which have approached $2 trillion in the past six years and now constitute approximately one-third of the U.S. federal budget, have come to dominate the economy. These funds have become a primary source of support for scientific research, and have deflected scientists and facilities from civilian to military purposes—a factor in the U.S. decline in advanced industrial technological innovation.

Even more relevant to the immediate subject, however, is the progressive weakening of civilian control over the autonomy of the military establishment in driving the competition with the Soviet Union. George Kistiakowsky testified eloquently about how [former President Dwight] Eisenhower came to feel increasing despair at his inability to control the Pentagon. Eisenhower observed the Pentagon driving up estimates of presumed Soviet capabilities in order to get larger appropriations, on which military careers and profits for military contractors depended. He enlisted Kistiakowsky's help in an unsuccessful effort to exercise control over the Strategic Air Command, whose inflation of targeting requirements led to overkill beyond reason.

The most crucial aspect of U.S.-Soviet relations is the military competition between the two countries. It is evident that the mindless increase in weapons programs, driven by parochial interests on both sides, has locked the two countries into a rising spiral, with its consequent tensions, apprehensions, and costs serving the interests of neither. This process has created an arms race that is the prime source of the world's insecurity. It is sometimes said that arms control has been tried and has failed to reduce weapons, or to prevent new weapons from taking the place of those that are limited. One major reason why this has appeared to be the case is that political leaders have felt it necessary to protect existing or planned weapons systems in which the military establishment had an interest, in order to forestall military opposition to the proposed agreements.

"The depth and seriousness of Gorbachev's concern over the dangers of nuclear war and the costs of the military competition cannot be doubted."

The Soviets obviously have also been far from rational in managing relations with the United States, and some of the same factors that affect U.S. policy have operated in the Soviet Union with equal or greater force. Soviet domestic politics, including in some periods factional conflicts, have been an important determinant of policy. Although Soviet military institutions are structurally different from their U.S. counterparts, rivalries among the military services, competition among the design bureaus, a persistent pattern of overinsurance in military affairs, the extensive military influence over civilian society, and, during certain periods, the relatively greater autonomy of the military establishment have resulted in short-sighted military decisions that did not serve Soviet interests. Misperceptions of the United States and the outside world generally stemmed from the same psychological mechanisms operating in the United States, magnified by ideological rigidities that have shaped the perceptions and expectations of the leadership, and compounded in the past by a parochial unfamiliarity with the outside world. . . .

[Mikhail] Gorbachev's reforms did not spring into existence suddenly, by some form of immaculate conception. They represent the continuation and maturation of a process that has extended over three decades, and reflect a growing awareness on the part of many—although still a minority—that the system developed by Stalin had become increasingly dysfunctional. And while one man and his personal qualities may now largely determine how much of the transformation is ultimately achieved, it should be understood that his efforts are part of a process that reflects a logic of necessity for moving the system to a new historical stage of development.

The details of Gorbachev's reform program are by now familiar. His political slogans are the equivalent of John Kennedy's "Let's get the country moving again!" One word he uses is *perestroika*, which can be translated as "restructuring" or better, "reformation"; and another, *glasnost*, meaning "openness" or "candor," has now entered the American lexicon. We shall concern ourselves here with the aspects of his reform program that most directly affect Soviet-American relations, and with the much debated questions about how Gorbachev's efforts should affect U.S. policy toward the Soviet Union.

Gorbachev and the Economy

The heart of Gorbachev's program is to modernize the Soviet economy. The economy he inherited was marked by declining growth rates approaching stagnation, low productivity, widespread corruption, and a massive and lethargic bureaucracy. Most important of all were the lag in the advanced technological sector of the economy behind that of all other industrial states, and a system that tended to discourage innovation and initiative. The starting point for Gorbachev was a recognition of the worldwide revolution in science and technology—an idea that has been discussed fitfully at lower levels in the Soviet Union for more than a decade. The new phase of the industrial revolution, he has said, centers on advanced technology, computers, electronics, robotics, genetic engineering, information processing, and automated production. In these areas the Soviet Union was falling farther and farther behind, and if this were not corrected, the country would find itself not even a second-class power in international competition. . . .

Gorbachev's foreign policy, and policies on defense and arms control, are subordinate to his concerns with the domestic economy. This is a period of turning inward for the Soviet Union. Gorbachev has insisted that his domestic priorities require tranquility abroad, and not adventurism. There is a general recognition in the Soviet Union that the activism of the 1970s under Brezhnev proved costly, first because it incurred economic liabilities the country cannot afford, and second because it damaged relations with the United States.

In what he has called the "new thinking" required in the world, Gorbachev has put primary emphasis on the need to avoid a nuclear catastrophe, and to establish a dialogue and mutual understanding with the United States. His secondary emphasis has been on global problems: the interdependence of the world economy, in which he sees the Soviet Union inextricably involved; the plight of the developing countries and their massive debt burden; and environmental concerns.

Seeking Dialogue

The time has come, he has said, for a turning point in U.S.-Soviet relations; for a sober, pragmatic reassessment of the relationship on both sides.

Despite the military influences he sees as dominating American policy, he believes the national interests of both countries can and should lead to an easing of the military competition, as is required by the economies of both countries. As he put it in a colorful image, we should not be like "two dinosaurs circling each other in the sands of nuclear confrontation."

Remarkably enough, the Soviet Union has sought doggedly to engage the United States in productive dialogue and negotiations, as unpromising as this prospect has appeared to them on repeated occasions. Although not wanting to take the role of supplicants, the Soviet leadership has persisted in seeking even modest agreements with the Reagan administration, if more comprehensive agreements prove unattainable. Their hardheaded calculation is that the situation cannot be allowed to drift while new weapons systems are developed on both sides, and the Soviet Union would be obliged to continue to divert into the military sector resources sorely needed for the modernization of Soviet industrial technology.

"The confrontationist view that has dominated U.S. policy toward the Soviet Union is based upon assumptions that were questionable to start with."

The depth and seriousness of Gorbachev's concern over the dangers of nuclear war and the costs of the military competition cannot be doubted. The Soviet Union is more serious about its interest in engaging the United States in productive negotiations on arms control than ever before in history. But it is also clear that Gorbachev is walking a tightrope. He cannot afford to have the Soviet Union appear weak or intimidated by pressure; but he also feels the need to avoid taking actions that would fuel the notion abroad of a renewed "Soviet threat." This calls for a carefully calibrated policy, which has so far been evident. One can only speculate about the domestic pressure he must take into account. While some among the Soviet military leaders appear to support his policies, on the ground that a strong industrial base is necessary for the future of Soviet power, there are hints that others are concerned about the effect of cuts in their services, stemming from arms control agreements or budgetary reductions.

Several elements in Gorbachev's "new thinking" bear watching. One is his emphasis upon the "mutuality of security." As he said at the 1986 party congress: "We can never be secure while the United States feels itself insecure." If this view were to become dominant in the Soviet Union, it would show more insight than either superpower has shown before and could improve the prospect for negotiations. He has also observed that the Soviet Union has no need of an external enemy, breaking with Stalin's reliance upon a "capitalist encirclement" to justify military programs. His questions about Soviet military requirements suggest he is beginning to follow logic to the concept of sufficiency, as the United States did during a brief, earlier period. In the broader aspects of foreign policy, he has recognized the interdependence of states in the world economy, and has accepted the implication that autarchy is impossible at a time when economic problems are international.

Hopeful Prospects

These are "tender shoots" that might wither or flower, depending upon internal and external circumstances. But what they suggest is the possibility that the Soviet Union may have moved a long way from the "two camp" doctrine of hostility toward a period when, if encouraged, it would recognize its own interest in playing a more constructive role in the international system and in international economic institutions. This too could have an important effect on the debate within the United States on the prospects for a more constructive relationship with the Soviet Union.

What follows if we look at the course of American-Soviet relations in this way?

First, it seems clear that the confrontationist view that has dominated U.S. policy toward the Soviet Union is based upon assumptions that were questionable to start with, and are increasingly inapplicable to the Soviet Union as it has been changing over the years. It is a view that can have no outcome other than the continuation of an unregulated nuclear military competition which will be an increasing danger to U.S. security and a continuing distraction from other important foreign policy problems, whose resolution becomes more difficult as long as the two great powers are locked into a confrontational relationship.

Second, while it is too soon for summary judgments about prospects for the success of Mikhail Gorbachev's efforts to reform the Soviet economy and modernize the Soviet system, the steps he has taken over the past two years present an opportunity to put the U.S.-Soviet relationship on a more sensible footing. In the first instance, the United States should, in its own interest, explore the possibilities now available to stabilize and moderate the deterrent balance between the two countries, in nuclear weapons and in conventional weapons and forces, and to open dialogues at many levels on bilateral and multilateral economic, regional, and political problems, holding open the possibility of cooperative action.

But U.S. policy should also be formulated from the outset on the basis of a long-term evolutionary purpose. If the initial negotiations are sucessful, they should be seen as opening the way to a long-term shift—perhaps over decades—in the balance between competition and cooperation, with the purpose of drawing the Soviet Union into a constructive involvement in both the international system and the international economy.

Third, Americans should recognize that if they cannot respond productively to the opportunities presented by the Soviet Union, because of the persistence of the fundamentalist ideological views that have dominated U.S. policy in the past, the consequences will be disadvantageous to U.S. interests:

- Strains in the Atlantic Alliance will become increasingly serious.
- Other foreign policy problems—regional, political, and economic—will become increasingly difficult.
- To the extent that U.S. actions have an effect upon Soviet developments—perhaps marginal, but nevertheless significant—Gorbachev, if he survives politically, is likely to be driven to carry forward his reform efforts in a climate of hostility. The mobilization this would entail would diminish the chances for easing the repressive aspects of the Soviet system and the likelihood of changes in the Soviet system and foreign policy that the United States would wish to see.

An Enlightened View

Fourth, many of the most urgent U.S. foreign policy problems—the arms race, the budget deficit, the trade imbalance, the large Third World indebtedness, the sources of conflict among the developing countries—can only be solved within an international framework, which is out of reach so long as the nation continues to be preoccupied and obsessed by the political and military competition with the Soviet Union, and gripped by a nationalist fervor that undermines international institutions.

Finally, the key to U.S. movement in these directions is in its domestic political life. What is needed is a strong, politically effective constituency that supports an enlightened view of U.S. security interest in moderating the military competition and relations with the Soviet Union, along with a return to the bipartisan spirit of internationalism that characterized the immediate postwar years.

Marshall D. Shulman is senior lecturer in international relations and a former director of Columbia University's W. Averell Harriman Institute for the Advanced Study of the Soviet Union.

"While Gorbachev is talking disarmament, the Soviet Union is continuing to arm dangerously."

Soviet Arms Control Efforts Are Not Sincere

William R. Van Cleave

In his annual report to the U.S. Congress for fiscal year 1985, [then] Secretary of Defense Caspar Weinberger observed that "the Soviet drive toward superiority has been particularly pronounced in the realm of strategic nuclear forces." Overall, Soviet nuclear programs, he said, "have undercut the stability of the nuclear balance and undermined the retaliatory effectiveness that was at the heart of our policy of deterrence." Earlier, addressing the nuclear balance, President Reagan publicly acknowledged, with unprecedented candor: "The truth of the matter is that on balance the Soviet Union does have a definite margin of superiority—enough so there is risk."

These are highly unsettling statements. Given the centrality and criticality of the U.S.-Soviet nuclear balance to the relations between the two powers, to the military balance, and to the global and regional "correlations of forces" (to use a Soviet operational term), these conclusions must be treated with the utmost seriousness and importance. The U.S.-Soviet nuclear balance is critical not only because of the grave threat and possible consequences of nuclear war, but also because a nuclear imbalance favoring the Soviet Union magnifies force imbalances and deficiences below the nuclear level. Soviet superiority at the nuclear level not only undermines the credibility of the U.S. deterrent, it also erodes the deterrent capability of U.S. conventional military forces. If this is true for the United States, it is even more true for other powers confronting or threatened by the Soviet Union. Moreover, a nuclear imbalance, to the degree that it is in the Soviet favor, may well encourage the Soviets to be politically more assertive and militarily more adventurous, in confidence that other powers may back away from

William R. Van Cleave, "The US-Soviet Nuclear Balance: A Summary," *Global Affairs*, Fall 1987. Reprinted with permission.

confrontations having ultimate nuclear overtones. (Even should such an expectation prove false, it could lead the Soviets to create a situation that should have been deterred in the first place by manifest countervailing power. Such an expectation would pose a grave danger for peace and security.)

Glasnost's Goals

Glasnost has not changed this situation. It may well be that glasnost is a combination of trying to improve the Soviet economy (without changing the system) and seeking some relaxation of international tension (read: Western competitiveness and rearmament) in order to gain the time necessary for those improvements to bear fruit. But nothing has changed in the fundamental world outlook of the Soviet Union, in its foreign policy, in its inherent militarism, or in Soviet international opportunism. The Soviets can and must be expected to continue to practice a policy of intimidation, destabilization, and even opportunistic expansionism. They will not relinquish the military superiority that they have worked so hard to gain, whatever the apparent promise of General Secretary Mikhail Gorbachev's arms control posturing. In fact, there is every indication that Soviet military, particularly nuclear, programs will continue to expand. And the Soviets still believe nuclear superiority to be the *sine qua non* for the type of military and political advantage over the West necessary to the achievement of Soviet international goals.

Correcting the nuclear imbalance, then, deserves the highest priority on the part of the United States. . . .

A study conducted in 1987 for the Department of Defense identified more than forty categories or indexes of relative U.S.-Soviet strategic nuclear strength and traced those comparisons from 1962 to 1982. In 1962, all favored the United States; in 1982, virtually all favored the Soviet Union. This profound change over a period of twenty years demonstrates conclusively the disparity of efforts between the two states and the fact that there has not been during that time what could reasonably be called a "nuclear arms race." Had the United States been racing such a complete change could not have occurred.

"Despite the awesome Soviet strategic offensive nuclear buildup . . . of recent years, there is no indication of any slackening in this buildup in the future."

Moreover, in contrast with popular belief, while the Soviet nuclear weapons stockpile has been growing dramatically, the U.S.-nuclear stockpile has been unilaterally reduced. Compared with the 1960s, the U.S. nuclear stockpile has about one-third fewer weapons and only one-fifth to one-fourth of the total yield or megatonnage. This has been a unilateral process, not required by any arms control agreements and certainly not reciprocated by the Soviet Union. On a multilateral alliance basis, NATO decisions, taken without regard to the ongoing arms negotiations with the Soviet Union or the continuing buildup of Soviet theater nuclear forces, have reduced in recent years NATO theater nuclear forces by 2,400 weapons. It is ironic that Gorbachev gets more credit simply for talking about possible reductions than NATO leaders get for actually reducing nuclear forces.

A Widening Gap

The Soviet Union first exceeded U.S. annual military spending in 1969, and the gap between the two steadily widened. Between 1972 and 1981, U.S. military spending actually declined at an average annual real rate of 2 to 2.5 percent, falling to under 5 percent of gross national product (GNP). By the end of Reagan's first term, U.S. spending had risen somewhat to 6.2 percent of GNP, but in the past two years there has been no real growth, and even if the current five-year defense plan were fully funded (which is unlikely given the present political situation) the figure would again be under 6 percent of GNP. Of U.S. defense spending, by far the largest portion goes to manpower, and by far the smallest portion goes to nuclear forces. Nuclear forces typically command only about 10 percent of the U.S. defense budget; strategic nuclear force programs, excluding SDI, were only 8 percent of the fiscal year 1987 defense budget.

Very nearly the opposite is true for the Soviet Union. Soviet military spending has increased at an impressive rate, growing to between 15 and 20 percent of GNP; most of the Soviet "defense" ruble goes to armaments production, and a heavy portion of Soviet military spending goes into nuclear weapons capabilities, which the Soviets hold to be central to the military balance. The U.S. economy is twice as productive and strong as the Soviet economy, at least according to Western calculations of gross national product. Even so, Secretary Weinberger informed Congress that if the United States devoted the same percentage of GNP as the Soviets, its defense budget for fiscal year 1988 would not be $300 billion but more than $700 billion. This indicates something of the difference in priorities and relative military effort between the two states.

According to American estimates, the cumulative difference in military spending between 1970 and 1985 was on the order of $750 billion, of which $500 [billion] was investment; the difference in spending on strategic nuclear forces was on the order of $350 billion, or over three times the total U.S. spending. . . .

Despite the awesome Soviet strategic offensive nuclear buildup and modernization of recent years, there is no indication of any slackening in this buildup in the future.

The CIA's Report

A 1985 unclassified CIA report to Congress concluded that "strategic forces will continue to command the highest resource priorities in the Soviet Union" and that:

> The Soviets are increasing their resource commitments to their already formidable strategic forces research, development, and deployment programs. We estimate that total investment and operating expenditures for projected Soviet strategic offensive forces and strategic defensive forces (assuming no widespread ABM deployments) will result in a growth in total Soviet strategic force expenditures of between 5 and 7 percent a year over the next five years. (The rate would be 7 to 10 percent if widespread ABM defenses were deployed.)

Since SALT II was signed in 1979, the number of warheads on SALT-accountable strategic force launchers in the Soviet Union has more than doubled, from about five thousand to some twelve thousand; and most of these have good counterforce capability against U.S. land-based retaliatory forces and communications. There are, in addition to SALT-accountable forces—i.e., those forces deployed on SALT-defined "launchers"—an unknown but undoubtedly large number of additional missiles and warheads for refire or for other possible means of launching. A comparison of numbers and capabilities of Soviet forces today with that provided for the 1979 SALT II data base shows substantial growth; a comparison of U.S. forces with the U.S. SALT II data base shows little change and, in fact, a reduction in numbers of strategic bombers and ICBM and SLBM launchers.

In addition to continuing the modernization and

deployment of the SS-18 and SS-19 ICBMs and modern ballistic missile submarine forces, the Soviets are proceeding with the deployment of a new family of strategic offensive systems: the mobile SS-25 ICBM (of which there are now over a hundred); the SS-N-23 SLBM; the Blackjack strategic bomber; and a new family of 3,000-kilometer range cruise missiles for ground, sea, and air vehicles. An even longer range and more advanced cruise missile is under active development.

U.S. strategic nuclear offensive forces are being modernized and expanded in certain respects: At least fifty ten-warhead Peacekeeper ICBMs will replace fifty Minuteman III three-warhead ICBMs; Trident submarines are being produced at the rate of one per year (however, three Poseidon submarines were retired to keep within SALT II limits, despite Soviet non-compliance with SALT II), and beginning in 1989 each new Trident submarine will be equipped with more capable D-5 SLBMs; part of the B-52 force is being equipped with air-launched cruise missiles (but part of it is also being taken out of the strategic role and reassigned to conventional bombing missions); the first of an ultimate one hundred B-1B bombers are now operational; and improvements are being made in command-control-communications (C³) for the strategic forces. The mobile small ICBM remains under development with future deployment still tentatively planned, but its future is uncertain since it has fallen into disfavor within the Office of Secretary of Defense. The Pentagon places more emphasis on an additional fifty Peacekeeper missiles in a rail-garrison deployment mode that it does on the small ICBM, but the former seem unlikely to gain congressional approval for deployment.

"The Soviets know very well how to play the arms control process in order to take advantage of Western desires for arms control."

These are significant improvements, but they fall far short of the scale and scope of past, present, and planned future Soviet force programs. An open question is whether current U.S. capabilities and currently planned programs will be adequate to reestablish "essential equivalence" with the USSR, to close the "window of vulnerability" for U.S. deterrent forces in a timely manner, and to meet the standards and requirements officially established for U.S. forces. The prestigious Committee on the Present Danger has concluded that more must be done:

Without a significantly increased and sustained effort, it is highly unlikely that U.S. strategic forces will meet the officially established requirements for the maintenance of stability and essential equivalence at any time in this decade or in the early 1990s. . . .

Theater Nuclear Forces

There is no doubt that the theater nuclear balance, in both the Far East and in Europe, heavily favors the Soviet Union at this time, both in the numbers and quality of systems available. In 1984, the NATO alliance concluded: "The Warsaw Pact shows a continuing build-up of their nuclear forces across the entire spectrum. In Europe, the Warsaw Pact has an advantage over NATO in all major categories of nuclear forces." The same assessment applies to the Far East. This extends, as well, to short-range tactical or battlefield nuclear forces. Former American advantage in short-range tactical nuclear forces (primarily based on nuclear artillery) has been overcome by the Soviet Union, which now holds the advantage in such systems. New nuclear-capable Soviet artillery—152mm, 203mm, and 240mm self-propelled mortars—are deployed in the European and Far East theaters.

The balance of theater range nuclear aircraft also favors the USSR in numbers and in capability, especially when differentials in defenses are considered. Few U.S. aircraft have intermediate ranges with the ability to penetrate Soviet defenses at low altitude, day or night, in any weather.

Given these Soviet advantages, it might be argued that arms control agreements eliminating them in certain categories and establishing equal levels near zero are better ways to achieve balance than attempting to build up forces (always difficult in the democratic societies of the West) in which the Soviets already enjoy large advantages. The SS-20 force, in particular, is a large, powerful, and threatening force that dominates the LRINF balance; it is a force uniquely suited to a surprise disarming attack. Eliminating this force, or even reducing it significantly, has been a goal that has preoccupied the alliance for nearly a decade. Trading some three hundred U.S. nuclear warheads for an SS-20 force four times (or perhaps ten times) that size certainly seems a far better bargain than any yet obtained from the Soviets in arms control negotiations.

Playing the Arms Control Game

Why then should the Soviets offer such a good deal? They never have before. In the first place, there are reasons to doubt the genuineness and aims of the Soviet proposal. The Soviets know very well how to play the arms control process in order to take advantage of Western desires for arms control, which lead to unilateral constraints on Western arms modernization well in advance of any actual Soviet reductions and even in advance of agreement. The Soviets are adroit at holding forth promises of future limits and reductions in return for present constraints on Western programs. The Gorbachev proposals may simply be another such gambit. As such, it is most directly aimed toward generating pressures in the West to limit SDI, to keep SDI within the limits of the ABM Treaty, and to prevent any deployment of ABM systems in either the United States or Europe.

"There is not much basis for reasonable confidence in, or optimism over, Soviet arms control proposals that seem designed mostly for political leverage."

It may be, however, that Gorbachev is willing to reduce a large Soviet advantage in a particular class of weapon in return for political and strategic advantages that more than compensate: contention and confusion in NATO; doubts among allies about U.S. security commitments, priorities, and nuclear assurances; the benefits to the USSR of a heightened "détente" atmosphere in the West; and self-imposed constraints on Western arms programs. This trade-off would not reduce, but could possibly increase, the overall Soviet military advantage. SS-20 target coverage is more than handled by the excess Soviet ICBM capability, including the current deployment of the mobile SS-25 ICBM, in a way an ICBM version of an improved SS-20. Shorter-range Soviet nuclear systems complement this capacity. On the other hand, removal of U.S. INF would eliminate the only real capability in Western Europe that could threaten important military targets deep in Eastern Europe and in the western Soviet Union. It would also eliminate a force that Europeans (and many Americans) believe to be essential to flexible response and that has assumed a huge political and psychological importance in Europe, well beyond its military value.

No Grounds for Optimism

In fact, while Gorbachev is talking disarmament, the Soviet Union is continuing to arm dangerously. Soviet military programs continue to expand Soviet nuclear capabilities, as well as conventional force, chemical, and biological warfare capabilities. While calling for elimination of the SS-20, the Soviets are proceeding with an even more capable follow-on to the SS-20 as well as deployment of the SS-25. So, there is not much basis for reasonable confidence in, or optimism over, Soviet arms control proposals that seem designed mostly for political leverage.

There are other problems as well. Experience informs us that agreements do not do what we expected of them at the time of their conclusion.

The SALT I agreement in practice, for example, was substantially different from the one the U.S. government thought it had negotiated. This is because the Soviets always insist upon ambiguities and loopholes, of which they then take advantage, and the United States accepts such looseness in order to have an "agreement."...

There are sound reasons to conclude that strategic defenses against nuclear weapons, including those delivered by ballistic missiles, are feasible—indeed, in the USSR, a wide variety of such defenses, including ballistic missile defense, exists today—and that technologies are shifting the offense-vs-defense equation in favor of ballistic missile defense. Given the magnitude of Soviet ballistic missile defense programs, which are far more long-standing than U.S. programs, in many ways more advanced, and in the past decade funded at a level of fifteen times U.S. funding, it is clear that the Soviets believe this to be true. So, clearly, does Reagan.

Ballistic missile defenses, as defenses against any type of threat, from invading armies to attacking aircraft, can be worthwhile and strategically valuable at various levels of effectiveness. No defenses are likely to be 100 percent effective, and such a standard should not be set for ballistic missile defense. Limited defenses, as one expert SDI study pointed out, can

> greatly complicate Soviet attack plans and reduce Soviet confidence in a successful outcome at various levels of conflict. . . . Even U.S. defense of a limited capability can deny Soviet planners confidence in their ability to destroy a sufficient set of military targets to satisfy enemy attack objectives, thereby strengthening deterrence. Intermediate defenses can also reduce damage if conflict occurs.

The goal of the U.S. SDI program is expressed as a range of effectiveness:

> An effective strategic defense would help deter attacks against us and, if it is as effective as we hope, virtually eliminate the terrible damage that would occur if deterrence fails. . . .

An Already Precarious Balance

Defense against ballistic missiles and improved defenses against aircraft, in both the United States and on the territory of our allies, could go a long way toward equalizing and stabilizing the U.S.-Soviet nuclear balance. On the other hand, continued Soviet deployment and modernization of such defenses, while the United States and its allies restrain themselves to research and development, could greatly extend Soviet nuclear superiority and destabilize even further the already adverse and precarious balance.

William R. Van Cleave directs the Center for Defense and Strategic Studies at Southwest Missouri State University. He is also a senior research fellow at the Hoover Institute at Stanford University.

"Arms control . . . has failed as a means of constraining the Soviet Union and succeeded as a means of constraining the U.S."

The Soviets Use Arms Control To Gain World Dominance

Malcolm Wallop and Angelo Codevilla

American liberals, either unaware of or ignoring the historical experience of democracies with arms control, pressed ahead. They did not recall how the German aero sport clubs of the 1930s became the Luftwaffe. They did not recall that with on-site inspection in the shipyards, the Scharnhorst and Gneisenau materialized from arms control cruisers into the largest battleships of their era. They did not recall that their European liberal counterparts of the 1930s had Mussolini dead to rights when in 1936 the new cruiser Gorizia was actually weighed and measured 10 percent in excess of the cruiser weight limits. The British Committee of the Interior under Chamberlain's government overlooked the breach because they were trying to seduce the Italian government into a rapprochement with the 1936 London Naval Treaty. Violations then, as violations now, were no diversion to the pursuit of the process.

With history comfortably out of mind, American liberals had been telling the country that there just *had* to be a better way to provide for our security than to earn it, precariously, by never-ending vigilance, toil, and even the commitment of blood. In 1961 they had established an agency to spend millions of dollars to pay a lot of bright people to search for such a way. Without doubt, these people would come up with ideas. Their contracts called for papers and conferences. They surely would propose schemes. But it was up to high officials, both elected and appointed, to sift these ideas and to tell the difference between schemes that would make arms control serve the cause of peace—something that had proved impossible in the interwar and postwar periods—and schemes that simply made it *look* as if we were achieving security, at the price of actual insecurity. . . .

The premise of arms control is that weapons are not tools dearly bought to accomplish ends for the sake of which foreign leaders are willing to kill and die, but rather, that they are expensive burdens that these leaders would prefer to shed—if only we would do likewise. Americans can apply this premise to the Soviet Union only by misinterpreting the Soviet Union and what it is after, or by a mental process that, while acknowledging the Soviet Union's antagonism, discounts it. In fact, the arms control process is based both on misinterpretation and on the discounting of Soviet policy.

Why does the Soviet Union arm itself so heavily? Is it because of a kind of quaint paranoia, an excessive concern lest it be invaded yet once more? Does the Soviet government merely want to ensure, as it repeatedly stated in Stalin's early days, "socialism in one country," or does it want to cast a shadow beyond its own borders—a shadow that would foster kindred regimes and cause uncongenial ones to wither? How long a shadow does it wish to cast? Is the casting of a shadow optional for the Soviet regime, or is it something it feels compelled to do to the extent it can? Are the quantity and quality of its current armament programs consistent with the Soviets' acceptance of "socialism in one country," or with the desire to cast the longest and most effective shadow possible?

A Convenient Misinterpretation

In the early 1960s, American advocates of arms control provided answers to these questions very different from the consensus of the 1950s. The Soviet Union no longer armed itself heavily because it was bent on bringing the world under socialism. It was suddenly seeking only a minimum of power to deter an attack from the U.S. Had the Soviets ever sought world socialism, they abandoned it when the Kennedy administration embarked on a program to build 1,000 Minuteman ICBMs [intercontinental

Malcolm Wallop and Angelo Codevilla, *The Arms Control Delusion.* San Francisco, CA: ICS Press, 1987. Reprinted with permission.

ballistic missiles] and over 600 submarine-launched missiles. "Conventional wisdom" has it that in the Cuban missile crisis, Soviet leaders facing hundreds of American missiles with 14 ICBMs of their own forever learned that they could never win a war against the U.S. on the highest level of violence, let alone a contest in technology and industry. Hence, though the Soviets could be expected to improve their strategic forces to make them respectable, the Soviet threat would manifest itself primarily through support of "wars of national liberation" in the "underdeveloped nations." Political competition would continue and would include violence, but it would take place in the back alleys of the world, unaffected by American and Soviet possession of big bombs and rockets.

"None of these economic shortcomings have stopped or even slowed the Soviet Union from building the world's most powerful military force."

This convenient misinterpretation occurred despite the Soviet leaders' continued reference—whenever they discussed support of wars of national liberation, or anything else for that matter—to the "correlation of forces." According to this concept, any given instance of the conflict between the imperialist camp and their own cannot help but be affected by all the forces at the disposal of either camp—be they nuclear, conventional, unconventional, economic, social, political, etc. How, then, could anyone believe that the Soviets would try to encircle and squeeze the West through the underdeveloped world without at some point gaining the ability to fight and win a war against the U.S. on the highest and most sophisticated level of violence of which the two sides are capable? Arms controllers conveniently ignored the fact that, since the 1940s, American policymakers had realized that our own nuclear superiority could "cover" the nonnuclear and even the nonmilitary operations of even distant allies. In the jargon of the 1960s, "extended deterrence" could be secured by "strategic nuclear superiority," because "strategic nuclear superiority" would supply "escalation dominance." Not surprisingly, the Soviets, for their part, saw the "covering" role of strategic superiority in precisely the same way. This had nothing to do with Marxism. It was and still is common sense. [Soviet leader Nikita] Khrushchev's doctrine of "peaceful coexistence" explicitly stated that growing Soviet power would somehow cover movements of national liberation. The regular pronouncements of such movements have long pointed to the overall power of the Soviet Union as

the indispensable foundation of their own success. . . .

Since the dawn of time, successful diplomats have lived by the rule that ascertaining the other party's intentions is the indispensable prerequisite for any negotiation. What is the other side after? What does it want from us in these negotiations? Wise diplomats have realized that they cannot change the other side's intentions, and that it would be self-deceptive to think otherwise. Indeed, wise diplomats must assume that the other side is entering into negotiations in order to facilitate its achievement of some purpose—to remove obstacles from its own path rather than put them there. Hence the irreducible need to find out what the other side wishes to accomplish.

No responsible American claimed to *know* why the Soviets were talking to us about arms control. After all, U.S. intelligence does not have access to Soviet leaders' deliberations. So the easy acceptance of the most soporific explanations of Soviet motives cannot be counted as misinterpretations. They were noninterpretations—castles of imagination built on imagination—that stood only because those who built them dismissed *a priori* the need to test their own assumptions.

A Mythical View of Soviet Intentions

Here, then, are the cards with which American arms controllers built their *a priori* explanation of Soviet intentions. The Soviet Union was having economic difficulties. Despite the fact that it had placed a satellite in orbit before the U.S. and had beaten the U.S. in missilery, the Soviet Union was technologically backward and knew it could not afford unrestrained military competition with the U.S. The Soviet Union, it is true, had once been highly dangerous. But Soviet leaders had matured. They desired only the benefits of ordinary international intercourse. They realized that the world had become interdependent, and the future of all nations' prosperity lay in the peaceful arts of technology, trade, and finance—not in military confrontation. The arms control process would allow the Soviet leaders to do something they wanted to do anyhow, but they needed international support in order to restrain the appetites of their own military-industrial complex.

The temptation to build this castle of imagination seems perennial. On March 5, 1987, President Reagan said that, while he personally realized the Soviets had not given up on their plans of world domination, it was also true that Gorbachev had many economic problems and even worse trouble with the Soviet bureaucracy. It was, therefore, urgent for Gorbachev to have an arms control agreement so that he could get on with modernizing his country. It would seem logical, however, that such circumstances would more ideally suit U.S.

pressures on the Soviets to return to compliance with old agreements rather than a rush toward new ones. Yet to demand proof of Soviet intentions seems to be outside the rules of the game.

But these cards would not stand examination—much less could they be built into an edifice that could withstand pressure. Of course the Soviet Union has been having terrible economic woes since 1918! The Soviet Communist Party considers most of its people's privations as the foundation of its own power. None of these economic shortcomings have stopped or even slowed the Soviet Union from building the world's most powerful military force. Yes, the Soviet leaders know they cannot afford unrestrained competition with the U.S. That is why *they* seek to restrain the U.S. But by what logic does this imply that the Soviet leaders are willing to restrain themselves, especially since Americans have exacted no price whatever for the Soviet Union's lack of reciprocal restraint? The Soviets know as well as anyone what the difference is between guns and butter. But they realize better than American arms controllers that superiority in guns can persuade those who own the butter to make a peace offering of it. The arms control process would indeed give to the supreme authorities of the U.S. and the USSR the ability to curb their military forces. But what might occur if Soviet political and military leaders made common cause with American arms controllers to restrain the U.S. military without any intention of restraining their own?

Even today the Joint Chiefs use a strange twist on the verification versus intentions issue. Though the Soviets are clearly in violation of major provisions of the ABM [antiballistic missile] Treaty, we should not opt out of it, they counsel the President, for the Soviets are in far better position to exploit their advantages gained through noncompliance than are we starting from our compliance. But the Joint Chiefs fail to be persuasive in their next step of reasoning. How, if the Soviets have achieved that superior position under the ageement that controls us but not them, do we enhance our safety by remaining bound to it?

"The arms control process itself is based on premises about Soviet intentions that are . . . demonstrably false."

Whenever such questions were asked about their premises, American arms controllers would fend them off by saying that the arms control process itself is the only way of finding the answers. Indeed, they saw the process not so much as a means of inquiry, but as the only means of *manufacturing* the right answer, because the process would teach the Soviet leaders sophistication—the table manners of nuclear weapons. So, American arms controllers turned on its head the tradition that dedicates the opening round of diplomatic negotiations to finding out the other party's intentions. This is because to question Soviet motives would be to underscore that the arms control process itself is based on premises about Soviet intentions that are either demonstrably false, or at best, undemonstrable. Instead, they focused the arms control process, including the preliminary negotiating rounds, on lecturing the Soviets about what their intentions ought to be and surely must be. . . .

Sokolovskii's Book

Marshal V.D. Sokolovskii's book, *Soviet Military Strategy*, had explained that the Soviet Union would have to achieve superiority in missile strike forces for purposes of deciding the war during its opening phase. The rate of commencement of Soviet silo construction certainly was consistent with Soviet military thought. Nevertheless, our "best and brightest" figured that the Soviets just had to look at nuclear war as a mutually annihilating spasm. Thus, our "best and brightest" saw arms control negotiations as an essential means of confirming the Soviet military in their own outlook. Besides, they argued, given the kinds of missiles the Soviets then had, the SS-7s and SS-8s, the Soviets could not reasonably try to destroy our missile silos. They concluded that the Soviets were not even technically able to choose for or against rationally waging nuclear war, and that therefore they had not done so. It was necessary for our "best and brightest's" peace of mind to suppose that the Soviets were technically backward and would remain so, and that the Soviets would wait for a technical capability before even deciding whether they wanted it! . . .

By 1967, however, fact came to trouble some of our arms controllers' dreams. The Soviets were testing a missile—that we called the SS-9—large enough to carry a 25-megaton bomb. Such a bomb, even given the SS-9's relative inaccuracy, was enough to destroy blast-resistant American silos. For most of American arms controllers, however, the SS-9 was easily explained away. [US Secretary of Defense Robert] McNamara and arms controllers rejected the notion that the SS-9's combination of huge yield with somewhat mitigated inaccuracy might be the Soviet Union's way of achieving the capability to fight and win a war. Instead, they called it an anti-city terror weapon. McNamara and his cohorts had to admit, however, that such concentrated high yield was a peculiarly inefficient way of destroying cities. Arrogantly, they resorted to the explanation that the Soviets had built such inefficient city-killers because they were technically unable to build efficient ones like the U.S. Poseidon. Nevertheless, beginning in the late 1960s, as the number of SS-9s grew and grew, even these Americans could not deny at least the possibility that the Soviet SS-9 missile—at least in two- or three-on-one attacks—could be used to destroy American Minuteman silos. Sadly, only after McNamara had left office, when the number of SS-9s passed the number of Minuteman launch control centers (100), did the argument about the SS-9's role change. Everyone agreed that great numbers of SS-9s *could* be a threat to the Minuteman force itself, and it became a national priority somehow to avoid their actually becoming one. Since roughly 1968-69, avoiding, and later fixing, "Minuteman vulnerability" became America's number one strategic obsession. Among the many options for dealing with this problem that the Soviets now posed was to ask the Soviets to limit it and solve our problem for us.

The SALT I Talks

So, by the time the SALT [strategic arms limitations talks] I talks formally began in late 1969, a totally new reason had arisen for seeking an arms control agreement. The U.S. now, for the first time, had a serious military problem on its hands. . . .

Given the intellectual inclinations and political priorities of most of those Americans involved in the arms control process, it was not clear at the outset of SALT what the talks could be about.

Any inquiry into the Soviets' political and military reasons for entering into the arms control process had to be excluded from the negotiations as inherently futile and disruptive. Yet discussion of our own purpose—the maintenance on both sides of relatively small and, above all, militarily irrational nuclear forces—could be (and was) vigorous in private. Were we to try to bring public pressure on the Soviets by "going public" with this line of argument, it would lead to Soviet rejection and would make arms control itself unpopular in the United States. What, then, could the actual talks be about? There was only one subject left: verification. . . .

"[Intelligence] would fill the yawning absence of any U.S. policy for enforcing compliance. . . . If the agreements were verifiable, *how bad could they be?"*

At the outset of the negotiations, in response to the most tentative of feelers, the Soviet Union had made crystal clear it would not give the U.S. any information—much less knowledge. It would provide neither "facts" nor the ability to check them out.

In 1951, the Soviets had agreed to a statement of principles on verification of disarmament that

Americans took as an agreement that there should be "unrestricted access without veto to all places as necessary for the purpose of verification." The Soviets, however, had reserved the right to decide what would and would not be necessary to verify their compliance. . . . The Soviets would be delighted to show us that they had destroyed the arms they undertook to destroy, the factories where arms production is banned by agreement, and the sites where—by agreement—only a certain number of weapons exist. To do more would be to establish "an international system of espionage." (It is interesting to note that in 1986 at Reykjavik, Mikhail Gorbachev took precisely the same position!) . . .

Verification Magic

Arms controllers used the U.S. intelligence community's ability to learn about some things that happen in the Soviet Union to perform a kind of magic: intelligence would erase the whole concern about the incompatibility between the Soviets' geopolitical goals and negotiating objectives and our own. It would fill the yawning absence of any U.S. policy for enforcing compliance, and indeed allay concern for the quality of any agreements. If the agreements were *verifiable*, how bad could they be? Verifiability would give us the chance to right any wrongs. Who could doubt we would do so? This magic, in turn, was achieved by another magic: the U.S. intelligence community let the few things that it knew—or thought it knew—stand for many more things of which it had literally no knowledge. In turn, and by the same token, the arms control process then profoundly affected U.S. intelligence. Let us see how all this occurred.

The advent of the U-2 high-altitude reconnaissance aircraft in 1956 and of the first SAMOS [satellite and missile observation system] photo-reconnaissance satellite in 1960, along with the expanding American signals intelligence network, proved to be a powerful catalyst in this magic.

The myth of the satellites' near-magical powers became unstoppable when they showed that, contrary to earlier belief, the Soviets had only 14 ICBMs on launch pads in 1962. Banishing the ghost of the missile gap had made possible the resolution of the Cuban missile crisis—which had itself been brought on by a discovery made by high-altitude cameras. Thus intelligence officers started to argue that this kind of knowledge, in the nuclear age, was enough to resolve crises. President Kennedy had brandished the photos and the Soviets had retreated. Both scholarly and popular literature abounded with assurances that the Soviets would never again expose themselves to such humiliations and that the existence of space-based cameras was more than enough to keep them scrupulously faithful to agreements. This obscured the fact that the Soviet Union had backed down not because of exposure or because of an axiomatic nuclear parity, but because of very real U.S. military superiority. Like the cat that, once burned, suffered cold rather than get near the stove, American officials learned the wrong lessons from the Cuban crisis.

A Disastrously Inaccurate Conclusion

In 1962, the Arms Control and Disarmament Agency sponsored a "summer study" on verification that produced an intellectual consensus—that although the Soviets could cheat on arms control agreements, they could not do so significantly before they were discovered. This was not so remarkable given the military and intelligence technology of 1962. The "threat of the day" consisted of primitive, pad-launched missiles. Each took months to deploy. During the time it would take the Soviets to deploy a large force, the satellites of the day—despite the limited amount of film on board—could be reasonably expected to find most of it in time for the U.S. to react.

"Production, even of things as large as ICBMs, takes place where eyes from space cannot see."

Disastrous errors often consist of unwarranted extrapolations of the truth. If one could reasonably have expected missiles and launchers to remain forever frozen in the technology of 1962, if that snapshot were engraved in eternity, then it would have made sense to base national defense policy on U.S.-Soviet agreements to limit the number of launchers/missiles, and to focus U.S. intelligence on observing their construction and operation. But, already in 1962, one could easily foresee that missiles would soon be far more easily handled, that launchers could be reloaded and movable, and that accuracy as well as numbers of warheads would be very important. But the "best and brightest," who became the ruling element in those years, wanted nothing so much as to stop the clock at 1962. Hence the consensus assumed that projected increases in photo resolution, film capacity, rate of recovery, and sensitivity of antennae would translate directly into earlier and even more certain verification of arms control. This consensus helped to make arms control into the very core of U.S. arms policy, which also came to have as its objective freezing the military situation of the mid-1960s; and, when that proved impossible, recreating it. . . .

Since SAMOS, the U.S. has orbited three generations of imaging satellites—the latest being the KH-11 series, first launched in late 1976. Their acuity has improved so that they can reportedly discern objects only inches in size. Their orbits have ranged from less than 90 miles to nearly 300 miles high, giving the satellites the theoretical ability to see anything within several hundred miles of their ground track.

These satellites, however, have some limitations that even high officials too often forget. First, not only can a satellite not "cover" or "image" all that it can theoretically see, it cannot even continuously cover the square kilometer directly on its ground track—a line that if drawn to scale on a large globe would be imperceptibly thin. Clearly, thousands of such satellites would be needed to do what former CIA [Central Intelligence Agency] Director Stansfield Turner comically claimed we do routinely: keep track of everything important on the face of the earth! But these satellites are so expensive that the U.S. has always had fewer than a handful in orbit at any one time. They routinely take pictures of sites we know to be interesting, and at most, take one low-quality picture per year of vast countries.

Such snapshots are useful. By looking at tanks, missiles, factories, movements of people, plantings of crops, etc., one might conceivably learn much about what things are available to a government, something about what that government is doing with them, and some hints about what it intends to do. Yet, many of the things that make for military or economic capability normally take place under roof, or are brought outdoors only when imaging satellites are not overhead. Production, even of things as large as ICBMs, takes place where eyes from space cannot see. Development of new weapons and stockpiling of old ones takes place under roof. Even big, outdoor things do not fall easily into the satellites' sights.

What the Satellites Cannot See

The best example of this came along in 1983: the Soviet radar at Krasnoyarsk. This huge installation is unmistakable—once a camera is pointed at it. But the intelligence community did not find it for many months. Yet finding the radar proved to be only the beginning of the controversy about what the Soviets intended by building it. Did they mean to violate the ABM Treaty? More important, do they mean to have an anti-missile defense? While the Soviets and their apologists claim innocence, and while most of the world cries foul, the pictures are silent. Pictures tell but little about intentions, and their tales are inherently ambiguous. Most activities, even if they take place outdoors, also may be observed only partially or with great difficulty. Of course, activities are inherently more ambiguous than things.

What, then, can—and cannot—the satellites see of Soviet strategic forces? They can see research laboratories with known affiliations and measure their size, and, from how busy they are or from new facilities constructed, conclude that they will turn out "a better model" of their standard product. But they cannot tell what is coming, when it is coming, how good it will be, in how many copies it is being produced, or how it will be used. Satellites can see test ranges. If the Soviets allowed them to, they could see missiles being prepared for tests. But, unlike bombers that sooner or later must fly and be seen, the Soviets do not have to show us missiles, and they don't. We do see preparations at test ranges, from which we gather some hints—perhaps useful—about the missiles to be tested.

Satellites see factories where missiles may be produced. But since the Soviets know when the satellites are coming, the satellites never get to count how many missiles come out, or where they go. Satellites see fields where silo launchers are dug, fitted out, and sometimes loaded with missile cannisters in plain view. If the silos are known, in plain view, and in good weather, the satellites can measure them rather accurately, and count them. But they never see all the silos loaded, and have not the foggiest notion whether or not any or all have been loaded or unloaded. Indeed, never seeing the missiles themselves, the analysts *suppose* but do not *know* where they are. . . .

Hiding the Missiles

As they go about their work, analysts—and, even more, collectors of technical intelligence—do not normally contemplate the possibility that the people whose images and signals they are looking at are aware that they are being observed and might act to shape what the intelligence collectors and analysts see. Few U.S. intelligence officials have drawn the right conclusion from the fact that the Soviet Union, soon after the launching of our intelligence satellites, figured out that satellites in sun-synchronous orbits took pictures, and that since the 1960s the Soviets have placed sensitive things under roof when these American satellites were scheduled to pass overhead. The run-of-the-mill explanation is that the Soviets are hiding things—and that is correct. But in fact, by hiding certain things and not others, the Soviets are able to shape our conclusions. . . .

"The U.S. military predicament in the 1980s is precisely the one we had sought to avoid by entering into the arms control process in the first place."

The upshot of all this is that American arms control negotiators built an artificial world in which the larger reality of Soviet military power, and the purposes for which it might be used, disappeared as if deliberately pushed out of focus, while the eyes of the American people were directed to tiny tidbits of facts fashioned through the concurrent, mutually reinforcing efforts of the arms controllers, the

intelligence community, and the Soviet Union itself. The American people were told that the arms control process controlled Soviet arms. But given criteria according to which new things could be "old" or heavy things "light," and in which key words like "rapid," had precisely no meaning it was never clear what was to be controlled—other than the thinking of Americans.

Anyone even slightly familiar with the limitations of U.S. intelligence who considers the artificial quantities that the arms control process substituted for Soviet military realities can see in them so many invitations to Soviet deception. As we shall see, the Soviets in fact appear to have taken advantage of at least some of these invitations. . . .

The Nightmare Becomes Real

Arms control, then, has failed as a means of constraining the Soviet Union and succeeded as a means of constraining the U.S. The figures published in the Defense Department's annual report, *Soviet Military Power*, tell much of the tale. As of 1986, the Soviet Union had *at least* 5,720 warheads on its SS-17s, -18s, and -19s, plus some 500 on 420 SS-11s, -13s, and -16s. These, as we have said, possess the combination of power and accuracy required for a high probability of success in attacks against missile silos and other "hard" targets.

That is more than enough to "cover" the 1,000-plus U.S. strategic "hard" targets with two warheads, and still leave the Soviet Union with a huge, land-based reserve force—never mind its sea-based missile force which, as of early 1986, carried some 2,700 warheads. In addition, the Soviet Union's force of 441 SS-20s could not just "cover" every militarily significant target in Western Europe with a fraction of its 1,323 warheads, but also deliver a large number directly to the U.S. How many, we do not know, since each SS-20 launcher has an undetermined number of missiles for refire. To this we must add the Soviet Union's 12 Yankee-class submarines (recently converted to carriers of modern cruise missiles), some 180 Bear and Bison bombers, and over 250 Backfire bombers. All this means, in sum, that a Soviet surprise attack on American forces would leave the United States with considerably fewer than 100 ICBM warheads, with perhaps 25-75 bombers, and with some 2,400 warheads aboard perhaps 15 submarines, while the Soviets would retain nearly all their forces. Given this new military balance, what could the U.S. do?

As Secretary of Defense Weinberger said in December 1984, the U.S. military predicament in the 1980s is precisely the one we had sought to avoid by entering into the arms control process in the first place. Actually, the predicament is much worse than anyone imagined at the outset of the process.

Moreover, the usefulness of American forces remaining after a Soviet first strike would be reduced by Soviet defenses, built largely in accordance with the ABM Treaty. The SA-12 anti-aircraft, anti-warhead system that could be deployed in some 1,000 mobile firing units equipped with perhaps 6,000 missiles could offer solid protection to any Soviet installations that were targeted by fewer than, say, a dozen warheads, and that the Soviet Union chose to defend with SA-12s. This weapon would thus significantly reduce the effect on the Soviet Union of a spread-out American retaliatory strike. Thus, the very existence of the SA-12, plus our ignorance of how many units would be concentrated to protect what targets, compel the U.S. to concentrate its retaliatory strike onto a relatively small number of high-priority targets in order to increase the chances of destroying them. But the effectiveness of such a strike would be reduced by the Soviets' network of nine large anti-missile, battle-management radars on the periphery of their populated zone, tied to the Flat Twin missile engagement radars and their associated SH-4 and SH-8 interceptors. This combination is openly deployed around Moscow, and who knows how many are in warehouses around the rest of the country? Even if this system succeeded in protecting only half these hypothetical targets-of-last-resort, the U.S. would have little to show for the total expenditure of its remaining strategic forces.

Nothing To Stop the Soviets

Then, of course, since 1980 the U.S. has expected the Soviet Union to test a high-energy laser weapon in space sometime in the late 1980s. No one knows how effective such a weapon will be against American missiles, or how many copies the Soviets would place in orbit during a crisis. But there is no doubt that, when added to the other measures the Soviets are preparing against American missiles, both prior to and after launch, even a small number of first-generation Soviet laser weapons could be effective. Under these conditions, *it was clear by the early 1980s that no American could devise any plan for employing those strategic forces conceivably surviving after a Soviet first strike—no plan that would allow the U.S. to accomplish any reasonable military objective whatever.*

Arms controllers are unwilling to say that, after a Soviet attack, they (or anyone they know) would recommend to the President that he launch the remaining U.S. forces. They are even less willing to specify what targets we should go after; whether, after a Soviet first strike, we should be trying to kill soldiers, sailors, or factory workers—i.e., whom we should be trying to kill, or why we should be killing them. Nevertheless, they profess confidence that because the Soviets could never be sure that a U.S. president would not irrationally launch the crippled remainder of U.S. forces, the Soviets would never seriously contemplate an attack. But since the late

1970s American arms controllers have made this argument with less and less conviction. Sometimes they still claim that thanks to their efforts, the Soviets have removed some 1,000 ICBMs from their silos and have removed 12 ballistic missile submarines from the line. When pressed, however, these advocates uncomfortably acknowledge that they have no idea what happened to these missiles that supposedly no longer threaten us, and are downright sheepish about the submarines' conversion into more formidable weapons—i.e., launchers of very modern cruise missiles. Nor do they mention the fact that the new ICBMs have four, six, or ten times the number of warheads of the ones they replaced. . . .

> *"[Arms control] advocates uncomfortably acknowledge that they have no idea what happened to these missiles that supposedly no longer threaten us."*

Advocates of arms control have succeeded brilliantly within the U.S. A glance at American strategic forces shows that these forces have stuck close to the logic of Mutual Assured Destruction, despite the consistent flow of American targeting policy away from MAD since 1974. The U.S. has not increased the number of its ICBM warheads since the early 1970s—2,100 in all—or their potency since the Mark 12-A program in the mid-1970s. The number of warheads aboard ballistic missile submarines has remained steady at 5,700 since the late 1970s. Because plans have lagged for substituting some of these warheads with ones carried by MX and Trident II missiles respectively, the U.S. still does not have the ability to threaten large numbers of Soviet "hard" targets on a time-urgent basis. As of 1987, the U.S. had no plans for acquiring a force able to do that.

Hence, although the U.S. has advanced the state of the art of strategic weaponry, and although the technology available to the U.S. would allow us to build weapons fit for whatever rational strategy we were to choose, in the 1980s the U.S. remained equipped with a force that fit no strategy at all—not even Mutual Assured Destruction. Given the Soviet Union's ability to reduce the stock of U.S. strategic weapons even before they are launched, and to defend against the ones that are, the U.S. in the 1980s could not even expect to execute MAD—to count on killing 25 percent, or any set percentage, of the Soviet population. There are no plans in the U.S. for preparing to increase Soviet casualties. Indeed, were the U.S. truly interested in preparing a MAD-consistent defense strategy to maximize Soviet casualties, we would be building enhanced-fallout warheads, and perhaps also warheads to disperse toxins and exotic diseases. That the U.S. is not—as, indeed, it should not be—building such horrible weapons is indicative of the fact that there is no true interest here in utilizing American weapons against the civilian population of the Soviet Union. The supposed strategy of MAD is mythical as well as immoral. . . .

The Mobile Missile Scam

Thinking of weaponry in terms of arms control has also made it difficult for U.S. intelligence to understand what Soviet weaponry it sees. We cannot here do justice to this complex subject. Suffice it to say, however, that focusing our intelligence on arms control has enabled the Soviets to hide major military developments behind the screen of minor violations of arms control agreements. The clearest example involves mobile missiles. The Soviet Union, knowing what our intelligence satellites see, has shown us fleeting glimpses that its SS-16 mobile ICBM program, banned by SALT II, continued— though not on a militarily significant scale. Because these discoveries are important for arms control, American intelligence for many years continued to think of the "mobile ICBM problem" as synonymous with the SS-16.

"Americans have precisely zero means for enforcing the terms of any agreement, good or bad."

At the same time, however, the Soviet Union was building 441 mobile launchers for the SS-20. Each SS-20 missile could carry a single warhead to intercontinental range, and each SS-20 launcher could launch SS-16s as well as SS-20s. These 441 mobile (SS-20, SS-16) launchers are also able to fire the newest Soviet missile, the SS-25, which also violates SALT II. But, as of 1986, the Soviets have built fewer than 100 launchers that are distinctively for the SS-25. These they show to our well-known satellites. In the past, these glimpses of "forbidden" activities led U.S. intelligence to identify the "mobile missile problem" with the distinctive SS-16 launcher. Today, these "forbidden" glimpses are leading U.S. intelligence to identify the "mobile missile problem" with the SS-25 launchers—whose number is small— rather than with the capacity of the 441 SS-20 launchers to fire a large but unknowable number of SS-25s, -20s, and -16s at the U.S. Intelligence is supposed to warn of danger, and to provide the knowledge needed to do something about it, but talk and resources devoted to this kind of treatment of

the "mobile missile problem" do the opposite. . . .

All of this is to say that in the 1980s, after twenty years of the arms control process, the U.S. is left with a radically worsened strategic situation, with an impaired ability to judge military developments at home and abroad, with a near-total reliance on arms control for our safety and independence, as well as with a growing realization that Americans have precisely zero means for enforcing the terms of any agreement, good or bad.

Malcolm Wallop is a Republican senator from Wyoming. Since 1979, he has been one of the Senate's official observers of US-Soviet arms control negotiations. Angelo Codevilla is a senior research fellow at the Hoover Institute at Stanford University and a former senior staff member of the US Senate Select Committee on Intelligence.

"The Soviet Union has agreed . . . to adopt each one of these [peace] measures. The U.S., citing the need to 'catch up with the Soviets,' has rejected each one."

The Soviets Do Not Seek World Dominance

Michio Kaku and Daniel Axelrod

The ultimate justification used by the war-fighters for counterforce, first strike weaponry is that the Soviets are expansionist and the "focus of evil" in the world. We need to build these weapons because they are pursuing the same course, and even pulling ahead. To the war-fighter, the bottom line is the belief that the Soviets want to dominate the world, and think they can fight and "win" a nuclear war. So no matter how hideous or unpleasant are the theories and weapons of war-fighting, we need them to offset Soviet superiority.

The crucial questions that demand answers are:
- Who is really ahead in pursuing Escalation Dominance and building a first strike capability to back it up?
- What about Soviet expansionism?
- Who spends more on weapons? . . .
- What about treaty violations? Can we trust the Soviets?
- What are the Soviets' long range goals in the world?

Who's Ahead?

The war-fighters have painted a dismal picture of U.S. forces. For the past twenty years, they argue, the Soviets have been fielding ever more accurate ICBMs [intercontinental ballistic missiles] which threaten our Minuteman bases with a limited counterforce attack. The U.S., however, because of its misguided MAD [mutual assured destruction] policy during the 1960s and 1970s, allowed the Soviets to catch up and then surpass us in the number of ICBMs. Pentagon charts all clearly indicate that the Soviets have taken the lead in throw-weight (how much "payload" or weight their missiles will lift) and megatonnage (the total explosive power of their warheads).

Michio Kaku and Daniel Axelrod, *To Win a Nuclear War: The Pentagon's Secret War Plans*. Boston, MA: South End Press, 1987. Reprinted with permission.

Upon examination, however, there is less to these charges than meets the eye. For example, the claims about numbers of ICBMs and total throw-weight and megatonnage are all true, but also somewhat irrelevant. The Soviet lead in land-based ICBMs (1,600 to 1,000 ICBMs) can also be viewed as the Achilles heel of the Soviet nuclear force. Land-based missiles would be easy to locate and destroy in a disarming first strike. The MX and the Trident II missiles are specifically designed to destroy Soviet land-based ICBMs. The U.S. voluntarily, and cannily, placed a ceiling on its land-based Minuteman missiles in order to build up the invulnerable leg of the triad, submarine-launched ballistic missiles. The Soviet "triad," by contrast, is overwhelmingly dependent on its vulnerable land leg.

This difference in force structure between the U.S. triad and the Soviets' heavy dependence on land-based ICBMs leads to an imbalance in first strike capability. Although estimates vary, an approximate calculation of present U.S. missile accuracy shows that roughly 80% of Soviet warheads (leaving about 1300) can be destroyed in a counterforce first strike launched by the U.S., whereas the Soviets could at best eliminate almost half of U.S. warheads. Neither side enjoys a totally disarming first strike capability now, but this situation could change as newer, more precise missiles are deployed in the 1990s. In this arena, especially when the MX and Trident II missiles are deployed, the Pentagon is closer to achieving first strike capability.

Shivers Up Your Spine

[Former] Secretary of Defense Caspar Weinberger appeared on television once to explain the overwhelming Soviet superiority in throw-weight, or the total amount that each side can throw into space. He had two scale models with him, one a slim, almost frail looking missile he called the Minuteman; the other was a huge, hideous looking

Soviet SS-18 missile.

"Shivers went up your spine when you saw these two missiles side by side. The superior size of the Soviet missile was ominous," one commentator recalled.

Upon careful examination, however, this Soviet "strength" actually conceals weakness. Stacking missiles side by side is, crudely speaking, like comparing a catapult, which can hurl massive boulders at the enemy, to a machine gun capable of firing only tiny bullets. The contrast in superficial appearances belies the true military potential of these weapons. Likewise, Weinberger's disingenuous presentation obscures the essential fact that Soviet ICBMs need to be large in order to compensate for their relatively low accuracy and reliability. Similarly, the war-fighters rail against the latest Soviet Typhoon subs, which are much larger than the Poseidon submarine. This, however, says more about the Soviets' inability to streamline their weapons than about alleged Soviet superiority. Soviet subs carry far fewer missiles than do U.S. subs, and remain notably noisier and hence easier to track and destroy.

"[The American hawks'] estimate of the Soviet threat has often exaggerated or distorted the facts."

A similar relationship prevails in regard to Soviet warheads, which have greater explosive power, but are also less accurate than U.S. weapons. The Pentagon has deliberately striven for smaller, more accurate counterforce weapons. The Soviets have been less successful because of their technological inferiority. More megatonnage is testimony to Soviet weakness, not strength. . . .

The war-fighters' claims about Soviet superiority are misleading, largely based upon isolating secondary factors and ignoring the rest. Former CIA official Arthur Cox has noted the war-fighters' one-sided use of the facts:

> The American hawks are now in power, more so than ever before, most of them having found nests in the Reagan national security apparatus. Almost without exception, they are cold warriors and advocates of military superiority over the Soviet Union. Through the years their assessment of Soviet power and intentions has had an important influence on U.S. foreign and defense policy. Their estimate of the Soviet threat has often exaggerated or distorted the facts. Sometimes the evidence has been presented to the public in a deceptive manner.

There is nuclear "parity" between the two superpowers only in one respect: the ability of each superpower to "kill" the other a hundred times over. In practically every militarily significant area,

however, the U.S. maintains a substantial lead. There is no doubt, of course, that the Soviets are gradually catching up in certain areas, such as missile accuracy, solid-fuel boosters, the quality of their navy, etc. Nevertheless, according to official Pentagon reports, the U.S. has a considerable margin of superiority over the Soviet Union, especially in the key areas involving high technology, and that margin is widening, not narrowing. . . .

What About Soviet Expansionism?

The war-fighters describe the Soviets as a giant steel tank, poised to roll wherever the West shows weakness. Since Western forces are no match for the Soviets on the ground, the war-fighters claim, the only thing that stops the Russian juggernaut from expanding is the U.S. threat to use nuclear weapons in response to Soviet aggression. Alternatively, some critics of nuclear weaponry advocate increasing the Pentagon's conventional forces instead of relying on nuclear weapons.

The concentration of Soviet conventional strength is in Eastern Europe, but the Soviet military presence there does not overwhelm NATO forces. In fact, the size of the Warsaw Pact force is much smaller than that which would be necessary to overrun the equally formidable NATO forces. Even without nuclear weapons, the forces of the East and West are largely stalemated, with neither side possessing the conventional capability to successfully invade the other. . . .

In summary, although the Soviet Bloc maintains a formidable conventional force in Eastern Europe, it nowhere approaches the size or conditions necessary to overwhelm the West. As George Kennan, former Ambassador to the U.S.S.R., has pointed out, from a military point of view Europe is the *least* likely place where the Soviets would launch an attack on the West. . . .

Third World Revolutions

The war-fighters often argue that their main fear is not a direct Soviet surprise attack on the U.S. or NATO, which is unlikely, but the more insidious danger of "Soviet subversion." Subversion is a new kind of expansionism: a sinister, behind-the-scenes takeover under the guise of a Third World revolution. They claim that such revolutions—in Cuba, Vietnam, Nicaragua, etc.—are bad enough in themselves, with their nationalist and Marxist ideologies and their expropriation of American assets. But combined with their predilection for supporting one another and receiving support from a center—the Soviet Union—they form a worldwide front seen as both threatening to "U.S. interests" and difficult to control by conventional military means.

Certain areas of the world are even proscribed from any dealings whatsoever with the forbidden sphere of the Evil Empire. While appealing for more military aid to counter-revolutionary forces in Nicaragua in March, 1986, President Reagan clearly summed up this view: "Ask yourselves what in the world are Soviets, East Germans, Bulgarians, North Koreans, Cubans, and terrorists from the PLO [Palestine Liberation Organization] and the Red Brigades, doing in our hemisphere, camped on our doorstep." Likewise, a 1981 State Department White Paper claimed that the war in El Salvador was a "textbook case of indirect aggression by Communist powers."

"Even the CIA and the Defense Department agree that the Soviet Union has a clean record in adhering to its nuclear agreements."

The Soviets, of course, do not start all these revolutions. They start as rural and urban poor living under intolerable conditions try to overthrow an oppressive dictatorship (which happens, more often than not, to be backed by the U.S.). But the Soviets often do offer substantial economic, political, and military assistance, usually in the form of outdated but workable military equipment.

With the aid of the "Soviet-backed" network of states, impoverished revolutionary governments are no longer tied to the U.S. by aid (although they often request it anyway), and no longer need to rely solely on their own meager resources for defense against CIA-sponsored counterrevolutions. Some analysts argue that the mere existence of an alternative means of survival outside the U.S. sphere is the most threatening factor to the war-fighters. Furthermore, the purpose of Escalation Dominance is precisely to assert military power against these revolutions and prevent the Soviets and their circle of allies from helping to sustain them.

The war-fighters ignore the fact that the Soviets have been notably unsuccessful at controlling revolutionary governments (for example, China). They usually gain little direct advantage from backing a revolutionary group whose main goal was to kick out foreign domination in the first place. Occasionally, they pick up a military advantage (such as their use of the old U.S. base at Cam Ranh Bay in Vietnam) which helps them break the circle of containment. But mainly, the Soviets appear to seek the indirect political and ideological benefits to be reaped from an expanded circle of influence, rather than to seek the "encirclement" and "containment" of the U.S.

At one of Reagan's press conferences, the President presented charts and diagrams which clearly showed that the Soviets outspent the U.S. as a percent of gross national product [GNP]. If the Soviets are such "peace-loving" people, then why have they converted their nation into an armed camp? Why have they chosen to gut their economy in order to produce weapons of mass destruction?

The large Soviet military budget, however, perhaps says more about inefficiencies in their economy and their desire for national survival than it demonstrates their "warlike character." Even the most pacific nation, faced with an enemy building first strike weapons, would be under intense pressure to play catch-up. The Soviet Union is not, of course, the most pacific nation, but neither is it the most militaristic. And given previous Pentagon plans to destroy the Soviet Union with a first strike, perhaps one can understand, if not excuse, some of their paranoia and anxiety.

More importantly, the main reason the Soviets spend a larger fraction of their GNP on arms is because their efforts to match U.S. capabilities draw on a much smaller economy. The gross national product of the Soviet Union is about half that of the United States, which means that a greater percent of their resources must be allocated simply to keep up with the United States.

Economic Games

Traditionally, the war-fighters' economic strategy has been to "spend Russia into a depression." Knowing that their economy is significantly smaller than ours, the war-fighters have tried to force the Soviets into matching the most expensive U.S. nuclear weapons. They have been partly successfull at this. Given their position of relative inferiority and the U.S. drive to build first strike weaponry, the Soviets have felt for some decades that they simply cannot quit. They believe that their national survival depends on maintaining nuclear equality with the U.S., even if it means having fewer consumer goods.

In the long run, however, the strategy of "spending the U.S.S.R. into a depression" may backfire. In this game of economic "chicken," the Pentagon may be unable to force the Soviets to jump before they themselves go over the cliff. The Soviet planned economy can better withstand the enormous pressures caused by maintaining a huge military establishment. For example, the *UN Yearbook of National Account Statistics* in 1976 stated that the Soviet rate of economic growth for 1966 to 1975 was 2.4 times the U.S. growth rate. This period, it should be noticed, was marked by intense military spending on both sides, with the U.S. involved in Vietnam and the Soviets trying to attain nuclear parity. In the 1980s, the Soviet growth rate dropped steeply before leveling off at equal the U.S. rate (which on the average is quite sluggish, at 2.5 to 3%). It is difficult to conclude that the Soviet economy is any more likely to collapse than that of the U.S. . . .

The bottom line, according to the war-fighters, is

that the Soviets simply cannot be trusted. It is desirable to have a commanding lead in counterforce weapons because "you can't trust the Russians." In particular, the war-fighters point to previous alleged instances of Soviet treaty violations. Reagan is fond of quoting Lenin's statement that "treaties are like pie crusts, made to be broken." (The quote from Lenin, however, is actually a statement denouncing the *Western* powers for breaking *their* treaties.)

Until the early 1980s, the question of treaty violations was not a major issue among experts in the "strategic community." Although both sides complained about "questionable activities" on the ambiguous fringes of the treaties, neither accused the other of major violations, even during the early Reagan years. For example, Roger Molander, formerly consultant to the National Security Council under Presidents Nixon, Ford, and Carter, has stated, " . . . the fact is, of the eight distinctive nuclear weapons-related arms control agreements signed by the Soviets in the nuclear age, there has not been a single charge of a Soviet treaty violation by the U.S., or vice versa. . . ."

"Missile warheads are difficult to verify because they are small and can be concealed—but a warhead is useless without a delivery system, and these systems are among the easiest to verify."

In fact, even the CIA and the Defense Department agree that the Soviet Union has a clean record in adhering to its nuclear agreements. In 1980, the Dept. of Defense, the Joint Chiefs of Staff, the State Dept., and the Arms Control and Disarmament Agency, stated that "Soviet compliance performance under 14 arms control agreements has been good."

Since 1972, questions about compliance with treaties have been addressed by a neutral body, called the SALT [strategic arms limitation talks] Standing Consultative Committee. The body meets every 6 months and has handled and resolved about 20 alleged violations of SALT I, mostly minor and technical, raised by both the United States and the Soviet Union. The creation of this body, one of the accomplishments of the Nixon administration, was designed to allow rational, careful study of charges of treaty violations by experts rather than permit a propaganda circus dominated by lurid, unproven charges of cheating by either side.

In the past, the allegations of SALT violations that occasionally surfaced in the press were mainly fueled by small right-wing organizations whose avowed object is to prevent any arms control treaties from being signed. Their allegations—that the Soviet

Union is testing the SS-16, that it has stored a number of ICBMs on top of one another in each missile silo, etc.—have never been accepted by the SALT Standing Consultative Committee. There is no small irony in the right-wing accusations of Soviet violations of SALT II, considering that the right wing helped to defeat SALT II in the Senate.

An Irrelevant Question

From the standpoint of arms negotiations, however, the question of "can you trust the Russians?" is largely irrelevant. The SALT treaties were not signed because the superpowers trusted each other. On the contrary, they were signed because each superpower trusted its own means of verification—the elaborate spy satellite systems, for example, that make it possible to track and monitor all nuclear weapons even before they are deployed. "Eyes in the skies" like the KH "Keyhole" spy satellite and the Big Bird satellite have been able to meticulously map every corner of the Soviet Union with amazing accuracy.

Advances in optics and photography have made it possible to see details down to less than 6 inches from a distance of one hundred miles. This has made it possible to tell, in principle, which newspaper a person in Moscow is reading by analyzing the typeface of the newspaper masthead.

Nevertheless, certain objects, such as missile warheads, are difficult to verify because they are small and can be concealed—but a warhead is useless without a delivery system, and these systems are among the easiest to verify. In fact, since rockets must be tested over several thousand miles of trajectory in outer space, their characteristics and trajectories can be measured with pinpoint accuracy by radars and computers. The very fact that the superpowers at times raise extremely obscure and technical treaty violation allegations before the Standing Commission is testimony to the accuracy of the means of verification.

In winter 1985, just before the Geneva summit, however, the war-fighters bypassed the SALT Standing Consultative Committee, leaking sensational charges of Soviet treaty violations to the press. These charges concerned the Soviet building of a phased array radar system, the testing of a new ballistic missile, and the encoding of missile telemetry. This public airing of alleged violations clearly goes against the thrust of the SALT treaty, which created the Consultative Committee in order to prevent precisely this type of media grandstanding.

Though these Soviet activities may not technically violate the treaty, they are ominous because they illustrate how the superpowers are testing the ambiguous, "gray area" of the SALT and ABM [antiballistic missile] treaties. Throughout the 1970s, during detente, neither superpower exploited the loopholes in the various treaties that were signed. In the 1980s, however, with a race on to build counterforce weapons, both the U.S. and the Soviet Union are pushing the SALT and ABM agreements to their limits. . . .

Radar Systems

For example, the ABM Treaty of 1972 did ban advanced radar systems, called phased array systems, from being used to track enemy missiles traveling over the interior of one's territory in a nuclear war. The reason for this was that any nation with a phased array radar system covering its interior could conceivably be building an ABM tracking system, which was banned by the treaty. However, phased array radar systems can be allowed under certain conditions, e.g., when they are used to track satellites, not missiles, or if they are located on the borders of a nation and are oriented outward, which greatly reduces their effectiveness as part of an ABM tracking system.

Recently, the distinction between an "inward" and "outward" phased array system has become a crucial element in the propaganda war between the superpowers. The Soviet Union has created an installation in Siberia near Kransnoyarsk in the interior of the Soviet Union, which seems to point inward, yet the Soviets claim it is strictly for tracking satellites, which is allowed by the treaty. The U.S., meanwhile, is building the PAVE PAWS phased array radar system in Georgia and Texas, which can cover up to two-thirds of the U.S. homeland because of its wide angle. The Soviets complain that this amount of overlap is prohibited by the ABM treaty. The U.S. is also upgrading two radar installations—at Fylingdales in Yorkshire, England, and Thule, Greenland—in apparent violation of the ABM treaty, which bans "overhaul, repair, or conversion" of these systems. The U.S. claims that this overhaul, however, is allowed because these installations predate the ABM Treaty. Highly technical charges and counter-charges fly both ways, stimulated not by full frontal violations, which neither side has committed, but by a politically motivated desire to discredit the other side. . . .

"If I had been the Soviet secretary of defense, I'd have been worried as hell about the imbalance of force."

In the view of Freeman Dyson, physicist at the Institute for Advanced Studies at Princeton and consultant to the Pentagon, "Our choice is not between imperfect and perfect arms control agreements; it is between imperfect agreements and none at all. An agreement does not automatically lose its value as soon as it is violated." For Dyson the real stumbling block to arms control is not so much the obscure, minor charges of treaty violations, but the substantial issues of politics. Do the superpowers have the political will to go ahead with meaningful reductions? . . .

The Soviet Grand Strategy

To the war-fighters, however, the question of Soviet treaty violations is only part of a much larger strategic question. While U.S. strategists in the 1960s agonized over MAD, they claimed that the Soviet military was already examining the possibilities of winning a nuclear war. The most prominent spokesman for this position is Harvard professor and Reagan advisor Richard Pipes. . . .

Pipes writes that:

> Soviet doctrine . . . emphatically asserts that while an all-out nuclear war would indeed prove extremely destructive to both parties, its outcome would not be mutual suicide: the country better prepared for it and in possession of a superior strategy could win and emerge a viable society. . . . The strategic doctrine adopted by the U.S.S.R. over the past two decades calls for a policy diametrically opposite to that adopted in the United States by the predominant community of civilian strategists: not deterrence but victory, not sufficiency in weapons but superiority, not retaliation but offensive action.

Pipes writes extensively about what he calls the Soviet Grand Strategy, a ruthless scheme by which the Soviets will slowly subvert the West and, if necessary, fight and win a nuclear war to attain their ends. . . .

As further evidence, Pipes and other war-fighters cite a quote by Vassili Sokolovsky, Marshal of the Soviet Union, from his 1960 book *Soviet Military Strategy*, considered by experts the basic statement of Soviet strategy. He wrote, "Military strategy under the conditions of modern war becomes the strategy of deep rocket strikes . . . to effect a simultaneous defeat and destruction of the economic potential and armed forces throughout the enemy territory, thus accomplishing war aims within a short period of time." Sokolovsky seems to think, or at least to assert, that a nuclear war could be won and "war aims accomplished."

Soviet Military Strategy

President Eisenhower's science advisor, George Kistiakowsky, once commented on Sokolovsky's statement about "winning" a nuclear war.

> Very influential was a book in the 1960s on military strategy by the Marshal of the Soviet Union Vassili Sokolovsky who asserted that the Soviet Union was prepared to fight a nuclear war and would come out victorious. This assertion has been later quoted *ad nauseam* by the American hardliners as proof that the Soviets were planning a nuclear attack on the United States. The book, however, was written at a time when the Soviets had relatively few atom bombs and were virtually defenseless against an American strategic at-

tack. The assertion was clearly in the nature of a morale builder by the top military commander for his troops. About the same time Melvin Laird, then a Congressman, expressed in a book his equally optimistic conviction about American nuclear victory and argued about the advantages of a pre-emptive strategic strike.

Laird, of course, was later Nixon's first Secretary of Defense.

When Sokolovsky first made his celebrated statements, the Soviet Union was vastly inferior to the U.S. in all aspects of nuclear weaponry. It was clear to everyone, including the Soviets, that they were in no position to talk confidently about winning a nuclear war. Yet later editions repeat and even amplify them with patently false assertions of Soviet superiority.

Why then did Sokolovsky, presumably a knowledgeable and able official, say something he and everyone knew was absurd? The answer has been fairly obvious to Defense Department circles, if not to Pipes. The principle is, if one has only a little stick, talk loudly. For example, former Deputy Undersecretary of Defense Morton Halperin, later an aide to Henry Kissinger on the National Security Council, wrote that "the Soviets presumably felt that a very small force coupled with exaggerated statements could have almost as great . . . a political payoff as a much larger force." As Eugenia Osgood, Research Analyst in Soviet Affairs at the Library of Congress summarizes the discussion, "Among some analysts the Soviet belief in victory here seemed more a patriotic exhortation than a realistic forecast."

"Soviet leadership views the U.S. leadership in much the same way as the Committee on the Present Danger views the Soviets: hostile, expansionist, and hell-bent on dominating the world."

Even if this generous view is wrong and the more alarmist view of Pipes is correct, it is not as if the Soviet Union was the only superpower planning to win a nuclear war. Sokolovsky's statement was made at the same time that the U.S. Air Force secretly requested funding for a full first strike capability from Secretary [of Defense Robert] McNamara. McNamara once candidly stated,

> If I had been the Soviet secretary of defense, I'd have been worried as hell about the imbalance of force. And I would have been concerned that the United States was trying to build a first strike capability. . . . You put those two things together: a known force disadvantage that is large enough in itself to at least appear to support the view that the United States

was planning a first strike capability and, second, talk among U.S. personnel that that was the objective—it would have just scared the hell out of me!

This point of view emphasizes seeing things through the eyes of the Soviet Union, *where every nuclear weapon but their own is aimed at their heartland.*

Furthermore, as pointed out by physicist Freeman Dyson, many of the early Soviet military writings about how their nation will survive a nuclear war are misinterpreted in the West. The vague statements about surviving the horrors of war are based more upon a literary metaphor than a precise military strategy. Deep within Russian culture is the stirring image of how Mother Russia will overcome overwhelming odds and, in the sense of Tolstoy, survive. The Soviets use the concept of "winning" in the sense of the inevitable historic victory of socialism over capitalism, not some decisive attack with counterforce nuclear warheads.

A Moot Point

Whether the Soviet statements about "winning" a nuclear war were meant to be bluffs to prevent the U.S. from launching a first strike, patriotic exhortations, ideological platitudes about the eventual victory of socialism over capitalism, or, as Pipes claims, the dark designs of the Soviet leadership, the entire point became moot in 1967 when the Soviet military abandoned the concept of nuclear war as a continuation of politics. That year, the Soviet military journal *Red Star* editorially repudiated the idea that nuclear war has any political utility. In 1971, the impossibility of a nuclear victory became the official assessment of the Soviet state. In 1969, at the first meeting of the SALT talks, the Soviet delegation announced their government's view:

> Even in the event that one of the sides were the first to be subjected to an attack it would undoubtably retain the ability to inflict a retaliatory strike of crushing power. Thus, evidently, we all agree that war between our two countries would be disastrous for both sides. And it would be tantamount to suicide for the ones who decided to start such a war.

By contrast, U.S. military doctrine has continued to evolve in the opposite direction, from "denying the enemy victory" in the 1970s to winning a nuclear war in the 1980s. . . .

The Soviet view, in contrast to Pipes', is that it is the U.S., not the Soviet Union, which is gearing up for a disarming first strike. In the book, *Whence the Threat to Peace*, the Soviets claim that the Trident II and the MX missiles are intended for a nuclear first strike. In their view the MX is "intended for a steep build-up of the potential for a first pre-emptive nuclear strike . . . the [Trident II] missile will have practically the same combat capability as the MX ICBM, that is, it will be a first strike weapon."

As "Americanologist" G.A. Trofimenko has stated,

> From the standpoint of theory, all this represents

classic preparations for acquiring the capability to carry out a first strike. It is true that, according to what they say, the American leadership denies any intention of this kind, alleging that its position is a defensive one . . . in fact, however, it is virtually impossible to reconcile the position of counterforce . . . with the position of strategic defense. Counterforce presupposes a first strike. This is also an axiom of nuclear strategy.

Ironically, the Soviet leadership views the U.S. leadership in much the same way as the Committee on the Present Danger views the Soviets: hostile, expansionist, and hell-bent on dominating the world.

"The Soviets are actually quite cautious, even conservative in foreign policy."

One way to resolve whether the Soviets really think they can win a nuclear war is to consider their position on limited nuclear war. Everyone, the Soviets, Pentagon officials, and independent observers, agree that a global nuclear war today would be suicide. Since such a war cannot be won, any nation serious about winning a nuclear war will have to plan to win limited nuclear wars. However, Soviet military writings, unlike their U.S. counterparts, have always stressed the fact that nuclear wars cannot remain limited, but will quickly escalate to a mutually devastating scale. Soviet military writings, including Sokolovsky's works, do not extol the nuclear escalation ladder; they are not predicated on the possibility of limited nuclear engagements in which the enemy surrenders before unacceptable damage is inflicted on both sides.

For example, Sokolovsky says that "any armed [superpower] conflict will inevitably result in general nuclear war," and warns that "if a war is unleashed by the U.S., it will immediately be transformed onto the territory of the U.S." Major General V. Zemskov wrote in a Soviet military journal in 1969 that "the concept of limited use of nuclear weapons is a lie of the Pentagon. . . . A nuclear fire which has begun cannot be put out by anyone."

These writings stand in stark contrast to those of the war-fighters, who from the time of [Paul H.] Nitze's NSC-68, through Kissinger's landmark book in 1957, through NSDM-242 and PD-59 of the 1970s and the documents of today, have always searched for a way to tame the fury of nuclear weapons for use in limited wars.

Who Are the Russians?

The final question is: who are the Russians? Are they on a quest for world domination, as Richard Pipes claims?

An alternative view of the Soviet Union is given by former Ambassador to the Soviet Union George Kennan, who at one time in his distinguished career

was a hard-liner who deeply distrusted Soviet intentions. Kennan's early cables from Moscow to the Truman administration, in fact, originally provided the ideological basis to NSC-68 and containment.

According to Kennan, the Soviet leadership is a:

highly experienced, and very steady leadership, itself not given to rash or adventuristic policies. It commands, and is deeply involved with, a structure of power, and particularly a higher bureaucracy, that would not easily lend itself to policies of that nature. It faces serious internal problems, which constitute its main preoccupation. As this leadership looks abroad, it sees more dangers than inviting opportunities. Its reactions and purposes are therefore much more defensive than aggressive. It has no desire to start any major war, least of all a nuclear one. It fears and respects American military power even as it tries to match it, and hopes to avoid a conflict with it. Plotting an attack on Western Europe would be, in the circumstances, the last thing that would come into its head.

This picture of the Soviet Union is also the one offered by prominent Soviet dissidents Roy and Zhores Medvedev.

Preserving Their Borders

This view posits a Soviet Union deeply worried about the integrity of its borders. The leadership desires "buffer states" between it and the Western powers, which they view as being hostile and eager for opportunities to destabilize the Soviet Union.

According to Kennan, the Soviets are actually quite cautious, even conservative in foreign policy. Kennan once suggested that, ". . . Communist ideology does not envisage any use, on Soviet initiative, of the Soviet armed forces for actions outside the country. This thus leaves no room, by implication, for the unprovoked initiation of hostilities against another great power."

Kennan notes that what emerges from an examination of seven decades of Soviet history is the sense of a nation obsessed more with preserving its own borders than rampaging across the borders of other nations. Kennan points out that the Soviet Union has rarely committed its troops outside its own borders, and even then only when the situation involved a "buffer" nation on its periphery, such as Afghanistan or Hungary. . . .

In effect, what has appeared as "aggression" and "expansionism" to hawks in the West for the last four decades could also be viewed as a Soviet desire for secure borders. According to Dr. Dyson, "Russians, when they think of war, think of themselves not as warriors but as victims."

Unfortunately, Kennan observes that it is difficult to offer an accurate reflection of the Soviet Union's intentions without being branded as "soft" on the Russians . . . or worse. Kennan laments that:

in the early 1950s . . . the exaggerated image of the menacing Kremlin thirsting and plotting for world domination came in handy. There was, in any case, not a single administration in Washington, from that of Harry Truman on down, which, when confronted with the charge of being "soft on communism," however meaningless the phrase or weak the evidence, would not run for cover and take protective action.

Regardless of the internal pressures which shape Soviet foreign policy, one of the most important indicators of the intentions of the Soviet Union is the peace proposals it has made.

An Opening for Peace?

On the international agenda now, there are numerous proposals for slowing or reversing the arms race: a comprehensive test ban treaty, no first use of nuclear weapons, a bilateral nuclear freeze, and no militarization of space. Although UN resolutions are an imperfect indicator of political will, the vote tallies of the 39th General Assembly are revealing.

In each case, the vote was overwhelmingly lopsided, with 120 to 150 nations voting "yes" along with the Soviet Union and only 2 to 20 voting "no" with the U.S.

The Soviet Union has agreed, in principle, to adopt each one of these measures. The U.S., citing the need to "catch up to the Soviets," has rejected each one. At times, the U.S. has distinguished itself by its lonely opposition to certain disarmament provisions. In 1984, for instance, the U.S. cast the sole "no" vote on six disarmament resolutions. No Western European country followed the lead of the U.S. on these votes.

"If both the US and the Soviet Union are to eliminate nuclear weapons systems simultaneously in measured steps, then parity is maintained at every step of the proposal."

In 1985, the Soviet Union increased pressure on the United States by announcing a unilateral moratorium on nuclear testing, taking effect after Hiroshima Day, and proposed a 50% across the board cut in nuclear weaponry. This 50% cut, in effect, would be much more substantial than a nuclear freeze. Then on Jan. 15, 1986, the Soviets offered a sweeping proposal to rid the world of all nuclear weapons by the year 2000, and even dropped their previous insistence that British and French forces in NATO be counted as extensions of U.S. forces.

This Soviet proposal took the Reagan administration by surprise. With much fanfare, the Soviets announced a comprehensive three stage plan

in which all nuclear weapons could be abolished. According to Premier Gorbachev:

Step One. Within the next 5 to 8 years, the USSR and the USA will reduce by one half the nuclear weapons that can reach each other's territory. As for the remaining delivery vehicles of this kind, each side will retain no more than 6,000. . . . The first stage will include the adoption and implementation of a decision on the complete elimination of medium-range missiles of the USSR and the USA in the European zone—both ballistic and cruise missiles—as a first step towards ridding the European continent of nuclear weapons. . . .

Step Two. At this stage, which should start no later than 1990 and last for 5 to 7 years, the nuclear powers will begin to join the process of nuclear disarmament. To start with, they would pledge to freeze all their nuclear arms and not to have them on the territories of other countries . . . all nuclear powers will eliminate their tactical nuclear weapons, i.e., weapons having a range (or radius of action) of up to 1,000 kilometers. . . .

Step Three will begin no later than 1995. At this state the elimination of all remaining nuclear weapons will be completed. By the end of 1999 there will be no nuclear weapons on earth. A universal accord will be drawn up that such weapons should never again come into being. . . .

Of course, one can always say that the Soviet proposal to ban all nuclear weapons by the year 2000 is a propaganda trick. What's important to note, however, is that this proposal is totally inconsistent with the principles of Escalation Dominance. Since there is no direct political coercion that can be derived from this proposal, the Soviets are not here pursuing Escalation Dominance. However, their proposal is totally consistent with Escalation Parity. If both the U.S. and the Soviet Union are to eliminate nuclear weapons systems simultaneously in measured steps, then parity is maintained at every step of the proposal.

Michio Kaku is a nuclear physics professor at the Graduate Center at the State University of New York. He is a fellow of the American Physical Society and project director of the Campaign Against First Strike Weapons. Daniel Axelrod is an associate professor of physics at the University of Michigan at Ann Arbor. He is a frequent lecturer on the arms race.

"Cheating and deception are an integral part of Soviet arms control policy."

Soviet Cheating Prevents Effective Arms Control

National Security Record

The major U.S.-Soviet arms control agreements, SALT [strategic arms limitation talks] I and the ABM [antiballistic missile] Treaty, were signed in 1972. The SALT II negotiations began shortly thereafter and the "arms control process" was in full flower. But by the late 1970s, reality had set in and arms control enthusiasm began to fade. A growing number of Soviet violations and circumventions were defeating U.S. arms control goals and bringing into question Soviet arms control intentions.

During the early 1980s Soviet behavior under arms control agreements was carefully examined by President Reagan's General Advisory Committee on Arms Control and Disarmament (GAC) and subsequently by other government agencies. The GAC report of December 1983, followed by five more reports issued by the president himself, detailed a number of Soviet violations of arms control agreements.

The question of Soviet intentions became urgent in 1983 when a U.S. reconnaissance satellite discovered a large phased array radar under construction at Krasnoyarsk in the central USSR. A large ABM radar in that location is a clear violation of the ABM Treaty, yet Moscow was building it with the knowledge that it could not be hidden and sooner or later would be discovered. The question was, why would the Soviets commit such a flagrant violation?

To gain a better insight into Soviet thinking, the CIA [Central Intelligence Agency] commissioned an examination of Soviet intentions in violating agreements, as seen "through the eyes of former communist officials." Emigres and defectors who were in high- and mid-level positions in intelligence, the military and even on the central committees of the Soviet Union and the countries of Eastern Europe were interviewed. They were asked for factual information and to analyze Soviet intentions. . . .

The Importance of Ideology

It is widely accepted by U.S. policy makers, Western academics and many Western journalists that Soviet ideology—that is, Marxism-Leninism—is dead and discredited. Yet the former communist officials stress that communist ideology remains the heart of Soviet organization, strategy and tactics.

Marxism-Leninism is fully dedicated to the success of the world revolutionary movement. This includes the destruction of all competing forms of society and the installation in their place of dictatorships of the proletariat. The top Soviet leaders believe their ideology and adhere to it, despite Western disbelief. A recent example was the surprise contained in media reports of the 1985 Reagan-Gorbachev summit; surprise that Gorbachev sounded like a *Pravda* editorial. *The New York Times* reported "the evidence suggests that the man (Gorbachev) sincerely believes these things."

The former communist officials emphasized that the single greatest shortcoming in the ability of the West to understand and deal with Soviet strategy, including Soviet arms control approaches, is the failure to accept Marxism-Leninism for what it really is . . . the basis of Soviet behavior.

Soviet ideology is militant; preparing for war is the most important activity of the state. The Soviet economy exists only to support the state, which means to support the Party and the tools it uses to maintain and extend its control. The principal tool of the Communist Party is the use of force or the threat of the use of force, which is why the secret police and the military figure so prominently in Soviet strategy. The Party's most important goal is to shift the global correlation of forces decisively in favor of the Soviet Union.

National Security Record, "Why the Soviets Cheat on Arms Control Theories," September 1987. Reprinted with permission.

Soviet strategy is a revolutionary strategy that stresses the use of subversion to undermine the strength of Moscow's adversaries and to promote disintegration from within. Soviet military strength is used to hasten the decay, through intimidation and demoralization, of the opposition.

Most Western policy is based on the assumption that Soviet strategy is pragmatic and opportunistic, that Soviet leaders are merely engaging in traditional power politics, and that genuine change in the Soviet system is possible. Many in the West do not believe that the Soviets have a global master plan, the purpose of which is to cause world revolution and win the struggle for the balance of power, since there is none in the West. But the former communist officials contend that this is a major Western error. The very essence of Marxist-Leninist strategy, they hold, is long-range planning to advance the world communist movement and create a favorable balance of power for the Soviet Union. And Moscow intends to achieve these ends through a combination of military superiority, intimidation and arms control.

"The Soviets use arms control activities to assist their drive for military superiority."

Following Stalin's death in 1953, Soviet strategy was revised in an effort to revitalize the communist movement. The essence of this new strategy was "peaceful coexistence." In this strategy, the Soviet Union is portrayed as changing, drifting away from Marxism-Leninism, and consequently the Soviets are seen as a diminished threat to Western interests. The Soviets badly need economic, financial and technical assistance and use peaceful coexistence, or détente, as it is now called in the West, to get it.

Moscow's first strategic priority has been to achieve nuclear superiority, because of its importance both in the event of war and to intimidate opponents, a traditional Soviet *modus operandi*. The Soviets use the nuclear threat to intimidate what they call Western "realists" into opposing defenses against nuclear war, arguing that the sole alternative to a nuclear holocaust is to support peaceful coexistence. Yet peaceful co-existence is itself an example of Soviet deception. In Soviet semantics, peace is achieved only when all opposition to communism is destroyed, thus peaceful coexistence (détente) is a means of destroying the enemy from within.

The Role of Arms Control

In the early 1960s the Central Committee of the Communist Party of the USSR and the Soviet General Staff advocated arms control approaches to the West as a strategic undertaking that would serve Soviet military and intelligence purposes. Day-to-day direction of arms control operations was made the responsibility of the Operations Administration of the Soviet General Staff. This is a highly secure organization that is responsible for the development of war plans and new weapons requirements. Cheating and deception planning was handled by the Administration for Special State Interests. Other participants were State Security (KGB), Military Intelligence (GRU), and the General Staff units responsible for materiel, science and technology.

The International Department and the Propaganda Department of the Central Committee promote Soviet arms control policies. They sponsor communist front organizations such as the World Peace Council to advocate Soviet arms control policies, and use non-communist gatherings such as the Pugwash conferences to influence Western scientists and academics to support Soviet arms control goals.

The Soviets use arms control activities to assist their drive for military superiority, to gain intelligence on Western plans and capabilities, to obtain feedback on the success of Soviet deception practices, and to aggravate relations between the U.S. and its allies.

Arms control, as approached by the Soviet Union, is a strategic military and intelligence operation with the following objectives:

- to derail U.S. defense planning and weapons acquisition programs;
- to verify Soviet intelligence on U.S. military plans and capabilities;
- to learn more about U.S. intelligence capabilities;
- to aggravate relations among the Western allies;
- to create disagreements within Western societies;
- to mislead the West on Soviet strategic goals; and
- to gain economic and technical assistance for the USSR.

At the same time, arms control must not be allowed to affect Soviet military planning or to enable the West to learn more about Soviet capabilities and intentions. That is, arms control in the West is completely different from arms control as pursued by Moscow.

Cheating and Deception

In 1963, the Soviet Union sent special instructions to its East European satellites on how to weaken the West politically and militarily. Included were guidelines on cheating, deception and disinformation, which were presented as very important activities because they blind the enemy to Soviet strategic goals.

The instructions said a primary purpose of deception was to deny the West the ability to evaluate accurately Soviet defense and economic capabilities, and to minimize the ability of the West to exploit the weaknesses of the Soviet Union. Should Soviet cheating be identified, Soviet strategy is to deny guilt while blaming the other side. An example of this Soviet tactic was Nikita Khrushchev's response to questions concerning the Soviet decision to end the nuclear test moratorium of 1958-1961 and resume testing in September 1961.

It was the United States, Khrushchev said, that was the first to test atomic weapons and to use such weapons to kill people. Second, in resuming testing the Soviets were merely responding to a series of U.S. tests conducted three years earlier, prior to the moratorium. And third, Khrushchev cautioned the U.S. not to resume testing, because that would begin a new round in the arms race. This approach, so artfully stated by Khrushchev in 1962, can be seen today in Soviet protests against the Strategic Defense Initiative.

Soviet plans to counter U.S. complaints about their treaty violations and other deceptions begin well in advance of their actions. They set the stage through propaganda and active measures for Moscow to deny guilt and shift the blame to the West. Then as soon as their activities are discovered, they swing their counterattack into operation.

In the early 1960s the Soviets examined their military force projections into the 1980s to determine how to hide important activities. Certain high-priority military research and development programs were transferred to or organized within "civilian" scientific and technical centers, to hide them better. Programs specifically identified for such deception included the development of new chemical and biological weapons, the development of weapons for use in space, and weapons based on new physical principles, such as lasers.

Soviet Risk Assessment

The Soviets consider the possible risks of the discovery of arms control violations and circumventions side-by-side with the potential benefits. An example of Soviet risk assessment was presented to East European leaders when the proposal to build the Berlin wall was discussed. The possible risks were assessed by the Soviets as follows:
- the West might encourage demonstrations and stir up trouble in Eastern Europe;
- the U.S. might take retaliatory action against Cuba;
- the effort to get peaceful coexistence accepted in the West might be stopped; and
- Western economic and trade sanctions might be applied.

The Soviets decided to build the wall despite these risks, because in Moscow's view the risks were outweighed by the following benefits:
- the wall would stop the infiltration of intelligence agents into the East;

- it would stop the flow of skilled workers to the West; and
- it would provide a good test of the United States and the Western Alliance. If successful, the wall would demonstrate that the U.S. could not be counted upon to stand up for its allies in a difficult situation.

Similar risk assessments are conducted when Moscow develops positions for arms control negotiations or propaganda. For example, in discussing with its East European allies in 1962 the Soviet decision to agree to a ban of nuclear tests in the atmosphere, the Soviets assured the East Europeans that the test ban would not be allowed to interfere with Soviet progress in nuclear warhead development, and that Moscow would test nuclear weapons in the atmosphere, regardless of the treaty, when and if it was militarily necessary to do so. The Soviets believed they could test without getting caught, but even if they were caught, explained General of the Army A.I. Antonov, then chief of staff of the Warsaw Pact forces, it would be useful in providing the Soviets with important intelligence about Western detection capabilities.

After signing the SALT I and ABM agreements in 1972, Soviet General Secretary Leonid Brezhnev explained to East European leaders that he had signed the agreements because it was the Soviet assessment that they would strengthen Soviet military power and lead to a decisive shift in the correlation of forces in Moscow's favor.

The Soviets give great importance to their assessment of the political dimension. In the case of the limited test ban treaty that outlawed atmospheric tests, the Soviets reasoned that an agreement would benefit them by enhancing the "peace-loving" image they were trying to project, and that it would whet the appetite for arms control of those private groups and individuals—academics, much of the media, and anti-war activists—who can influence policy in the West.

"Why do the Soviets deliberately violate arms control treaties? . . . Because cheating and deception are an integral component of Soviet national policy."

There is a major difference between the U.S. and Soviet approaches to arms control. The U.S. tends to play down if not to discount the political dimension. But it is of major importance to the Soviets, who regard the "arms control process" as a significant element of political warfare.

The Soviets are in no hurry to achieve agreements. Arms control, as they view it, is the long-term process of disarming their adversaries. Moreover, they benefit as much from the process itself as from any results that may be achieved. Consider, for example, the substantial anti-defense and anti-anti-Soviet movements that have emerged in the West during the years of U.S.-Soviet arms control negotiations.

How blatantly the Soviets will cheat on arms control agreements depends on what their military needs are and what they believe they can get away with. They are careful to prevent their cheating from having an adverse effect on their strategy of peaceful coexistence and the peace-loving image they try hard to project. They are well aware of how difficult it is for the West to "prove" their cheating, and how effective their system of very tight state security is in preventing the West from verifying their actions with any degree of assurance. Western reluctance to react to Soviet violations means that even when the West does discover Soviet violations it can operate to the Soviets' advantage. When the U.S. fails to respond to Soviet violations, it reveals to the world the weakness of the United States.

The most startling finding reached early in the [CIA] study was the high degree of unanimity of opinion of the former communist officials and the extent to which their knowledge and experiences were mutually reinforcing. No conflicts in reporting events and no substantial differences of opinion were found among those who contributed to the study. The second surprise was the emphasis these former officials placed on understanding communist ideology, which they consider absolutely essential to the successful analysis of Soviet behavior and intentions. The importance of this finding is heightened by the Western propensity to operate as though Marxism-Leninism were a dead ideology.

Superficial Change

They also emphasized that Soviet arms control objectives are very different from those of the United States. Indeed, the differences are so great that it brings into question the usefulness of the arms control process as currently pursued by the West. The former officials stressed that cheating and deception are an integral part of Soviet arms control policy.

Western foreign policy experts are constantly seeking signs of change in the Soviet Union. Some believe it is happening today with Gorbachev's Glasnost "reforms." But the former communist officials emphasized the opposite: that Soviet goals and long-term strategy remain unchanged. The highly publicized "changes" that occasionally occur in the Soviet system are no more than superficial adjustments. The leadership does not allow such changes to affect the nature of the Soviet system or its goals. Indeed, none of the former communist officials sees any hope of changing basic Soviet goals.

The apparent disregard in the West for long-term Soviet goals and strategy, the effectiveness of Soviet propaganda and disinformation, the absence of long-range Western strategy to defeat communism, and the importance accorded Western private interests (such as trade and finance) over national security, all work to Moscow's advantage and militate against any real change in the Soviet approach to the West.

Why do the Soviets deliberately violate arms control treaties? According to former communist officials, it is in their nature; it is called for in their ideology; and because cheating and deception are an integral component of Soviet national policy. Don't expect them to change.

The National Security Record *is a monthly newsletter that reports to the US Congress on national security affairs. It is published by the Heritage Foundation, a conservative Washington think tank.*

"The accounts of Soviet arms control cheating appear to be much exaggerated."

Soviet Cheating Does Not Prevent Effective Arms Control

Gloria Duffy

Judging by media coverage and the position of the Reagan administration, the Soviet Union has been engaged in massive violations of its arms control commitments. In a series of four reports issued since early 1984, the United States has charged the USSR with violations of nearly every major arms control agreement to which it has been a party during the past twenty-five years: including, the 1972 antiballistic missile (ABM) treaty, the 1972 Interim Agreement on Offensive Weapons, the 1963 Limited Test Ban Treaty, the 1979 SALT [strategic arms limitation talks] II accord, the Biological and Toxin Weapons Convention of 1972, the 1925 Geneva Protocol limiting chemical weapons, the 1974 Threshold Test Ban Treaty, and the Final Act of the 1975 Helsinki Conference.

Like the reports of Mark Twain's death, the accounts of Soviet arms control cheating appear to be much exaggerated. A number of the United States' charges seem to be mitigated by analogous activities undertaken by the United States, by Soviet correction of the behavior at issue, by common understandings reached in the Standing Consultative Commission (SCC) in Geneva, or by ambiguous evidence in support of the United States position. . . .

The Big Three of Noncompliance

Out of the dozen or more charges vetted by the Reagan administration, three of them—the big three of Soviet noncompliance—cannot be lightly dismissed. In the clearest case, the Soviets have constructed a large, phased-array radar (LPHAR) near Krasnoyarsk, in south-central Siberia. The ABM treaty prohibits such radar for ABM purposes, and Article VI of the treaty specifies that LPHARs for permitted early warning use must be located on the periphery of a country's national territory and be oriented outward. The Soviet radar, detected by United States intelligence when already in an advanced stage of construction in the summer of 1983, is situated between 900 and 1,000 miles from the nearest border of the USSR, which is the coastline on the Gulf of Ob.

A second direct Soviet arms control violation is that country's testing and deployment of two "new types" of intercontinental ballistic missiles (ICBMs), in excess of the one new type permitted by the terms of the SALT II treaty. In 1982, the Soviets announced that the mobile, multiple-warhead SS-24 was their one permitted new type of ICBM under SALT II. The Soviets then proceeded in 1983 to develop and test the three-stage, single-warhead, solid-fueled, mobile SS-25 ICBM. While the Soviets claimed that the SS-25 was a permitted modernization of the SS-13 ICBM, the capabilities of the SS-25 apparently exceed those of the SS-13 by far more than the 5 percent limit in various parameters set by SALT II.

Two qualifications must be added to this charge of Soviet violation. While the USSR, like the United States, has made a public commitment not to undercut the terms of the agreement, SALT II was never ratified by the United States, calling into question the legal status of its provisions. There is also some doubt, because of the limitations of United States intelligence capabilities when the SS-13 was initially flight-tested in the 1960s, about the baseline parameters of the SS-13 which must serve as the standard for charging that the SS-25 exceeds the permitted modernization levels. Despite these ambiguities, the SS-25 is probably a violation of SALT II that should be taken seriously by the United States, especially in light of additional new ICBMs which the Soviets now have in their development pipeline.

The Soviets agreed in SALT II to eschew encryption of telemetry from ballistic missile test flights which would impede verification of provisions of SALT II. Since 1979 the USSR reportedly has gradually increased its encryption to over 90 percent of all telemetry from missile test flights. Once again, the United States complaint is not airtight. All the while, SALT II remains unratified. More importantly, the provision of the treaty specifying that neither side shall "impede" verification is ambiguous, leaving to subjective interpretation what level of encryption constitutes an impediment to verification. With the level of Soviet encryption reaching virtually 100 percent, the USSR definitely is impeding United States verification, forcing the latter to rely upon less direct and effective means of verifying Soviet compliance with SALT II provisions. . . .

Two Schools of Thought

If the Soviets are violating key provisions of SALT II and the ABM treaty, this raises fundamental questions about Soviet intentions as a party to arms control agreements. Why have the Soviets undertaken the actions that have called into question their adherence to these treaty provisions? Two alternative explanations for Soviet behavior have been offered in the current public debate surrounding Soviet arms control compliance. Each of these rationales relies on a different image of Soviet national security decision making, and each is unsatisfactory in explaining Soviet behavior.

The first explanation for Soviet actions, which might be called the Violations Policy school of thought, has been consistently espoused by the Reagan administration since its first months in office. This view holds that the Soviets have deliberately and premeditatedly violated arms control agreements with the United States in order to gain unilateral military advantage. This position was most succinctly stated by [former] Secretary of Defense Caspar Weinberger in a report on "Responding to Soviet Violations Policy" (RSVP) submitted to President Reagan on November 13, 1985:

> The Soviet Union has been violating with impunity its principle arms control agreements with the United States. From the beginning many felt that the Soviets used the arms control process to obscure their planned offensive buildup, weaving into the fabric of the SALT I and the ABM Treaty the loopholes and ambiguities that they would later rely on to becloud or extenuate their violations. That pattern of Soviet behavior continues to this day.

Any attempts to negotiate compliance issues with the Soviets, in this view, have failed. According to the Weinberger report, the Standing Consultative Commission is an "Orwellian memory-hole down which our concerns have been dumped like yesterday's trash." The intent of such a violations policy by the Soviets, in the view of Weinberger and others, has been to cause the United States to

constrain its strategic programs through agreements, while the USSR gained military superiority. Soviet violations, Weinberger asserted, have given the USSR an advantage which "makes it very difficult for the U.S. to maintain a deterrent balance with them."

The other major school of thought about the reasons for Soviet compliance behavior, propounded by most of the critics of the Reagan administration's position, explains Soviet behavior in terms of bureaucratic errors or dysfunctions in Soviet decision making and as the result of circumstantial factors related to the legal status of the agreements in question and to the context of United States-Soviet relations. The Circumstantial/Bureaucratic image of Soviet behavior regards Soviet noncompliance as far less threatening and suggests that the Soviets have in the main complied with agreements, at least until the end of the 1970s. . . .

Questions Raised by the Two Schools

Each of these assumptions about Soviet decision making raises questions. In the case of the Violations Policy school, the question is: Who in the Soviet Union would have made the decision to adopt and put into practice a coherent violations policy, and what would the motivation have been? In the case of the Circumstantial/Bureaucratic image, the central question is: Could there have been a gap between decisions made about military programs and the commitment of Soviet leaders to the SALT II and ABM treaties? Is it possible that the provisions of SALT II and the ABM treaties were not well understood by the Soviet military and political leadership?

"If the Soviets have a coherent policy of cheating, why have they chosen to cheat in ways which bring the USSR comparatively little military advantage?"

The images that both schools present of Soviet arms control compliance decision making seem improbable. In the case of the Violations Policy school, if the Soviets have a coherent policy of cheating, why have they chosen to cheat in ways which bring the USSR comparatively little military advantage, at the risk of alienating United States participation in the arms control process? The actions at issue in the United States charges of Soviet violations—despite Secretary Weinberger's position that the United States deterrent has been degraded—have not substantially improved the Soviet strategic position. If the Soviets had wanted to materially advance their strength, they could have violated the fractionation limits of SALT II by increasing the number of delivery vehicles on Soviet large throw-weight launchers, or exceeded numerical ceilings on the launchers themselves, or taken a variety of other key steps.

> *"The Soviets have bargained hard in arms negotiations, seeking to advance their military security through arms control in much the same way as has the United States."*

The Circumstantial/Bureaucratic position is correct in emphasizing the probable lack of internal institutional checks on arms control violations in the USSR. Although they do not always function perfectly, in the United States there are mechanisms ranging from congressional oversight, to legal counsel within various departments, to the functions of the investigative press which help to insure that United States military programs proceed in accordance with arms control commitments. In the Soviet Union, there is no institutionalized advocate within the Ministry of Defense, the Defense Council, the Politburo, or the society at large for compliance with the terms of arms control agreements. If the Politburo needed to request a determination about the compliance of a particular program with the terms of agreements, such an opinion would likely be rendered by counsel from the Ministry of Defense, who would advocate whatever positions were favored within the ministry.

The Circumstantial/Bureaucratic interpretation's basic tenet—that an unintentional gap could have occurred between the Soviet military's understanding of Soviet commitments and that of the Soviet political leadership—is unsatisfactory for two reasons. First, it is likely that the top levels of the Soviet military and the Soviet military-industrial sector understand the provisions of the SALT II and ABM treaties. The number of top Soviet military officials who were involved either directly in the SALT I or SALT II negotiations or had some input into the decision making process in Moscow is large. There was extremely close contact between the military in the Soviet Union and the arms control negotiating process. The top level of the military leadership cannot be unaware of the obligations taken on in these agreements. Second, and at least as important, the agreements could not have been reached without their assent. . . .

Third Perspective

Neither of these over-simplified models—the Violations Policy or the Circumstantial/Bureaucratic approach—adequately explains the three important instances of Soviet noncompliance. A more complex

third view of Soviet behavior, which incorporates some elements of each of the other two approaches, better accounts for Soviet actions. This third view lays blame for Soviet noncompliance on three factors: wishful thinking of the United States about Soviet interpretations of the terms of agreements; an inadequate United States response to Soviet efforts to test the limits of United States tolerance regarding Soviet compliance with their arms control commitments; and the poor climate of United States-Soviet relations since the late 1970s.

In this view, the Soviets have bargained hard in arms negotiations, seeking to advance their military security through arms control in much the same way as has the United States. In several instances, including provisions restricting heavy missile launchers in the SALT I Interim Agreement and the encryption provision of SALT II, the Soviets have indicated during the treaty negotiations that they have strong reservations about specific treaty language favored by the United States. In both of these cases, the Soviets indicated at the time the agreements were being negotiated that they interpreted the terms as permitting them to continue ongoing military programs considered vital to their security. The negotiating record of both SALT II and the interim agreement convey Soviet interpretations of their obligations under particular provisions which differ substantially from the understanding of the United States. The United States opted to accept ambiguity in the provisions of these accords in order to achieve agreement and then chose to rely upon its own interpretation of the meaning of the Soviet commitments, despite Soviet signals to the contrary.

Radar Installations

The case of the Abalakova radar may be a bit different. This radar is the final link in a Soviet nationwide system of six early-warning radars. This network was planned in the early 1970s, and the decisions to proceed with the final installation were probably made between 1978 and 1980. The Soviets have placed the last radar in the series in the location that is optimal both for economies in construction and efficiency of operation. The site near Krasnoyarsk lies just below the permafrost level in Siberia, allowing for both greater ease of construction and greater structural reliability of the radar than if it had been built closer to the northern periphery of the USSR. Placing the installation any farther north would have necessitated the building of two radars rather than one in order to achieve the same degree of coverage of potential United States launches from submarine-launched ballistic missiles (SLBMs) in the Arctic Ocean, using the number of faces standard for these Pechora-class radars.

The Soviets might originally have planned to build the radar closer to the periphery of the USSR; but through experience in building and operating the other five radars, the Soviets may have discovered limitations in the system. They may have realized that their radar systems were less capable in practice than on the drawing board, and they probably learned that building such massive installations in the Soviet Far North was a more expensive and difficult proposition than they had estimated. Faced with the prospect of having to build not one but two of these radars on permafrost, the USSR may have decided to proceed at the more interior location irrespective of ABM treaty limits.

In the mid-1970s, the Soviets built the first in the series of LPHARs near Pechora, just west of the Ural Mountains in the northeastern region of the Russian republic. While it is closer to the periphery of the USSR than the Krasnoyarsk radar it still lies nearly 300 miles from the Soviet coastline. The United States did not strongly protest the situating of this radar at the time. Based on this precedent, the Soviets may have surmised that presented with a fait accompli at Krasnoyarsk, the United States would not effectively remonstrate.

> *"There might have been a Soviet Compliancegate at the onset of the Gorbachev administration, at which point the slide toward noncompliance was halted."*

American wishful thinking in interpreting agreements with the Soviets and a Soviet tendency to test the limits of treaty provisions are ongoing patterns which still did not produce significant problems with Soviet compliance prior to the early 1980s. Soviet noncompliance at the level of the Abalakova radar, encryption, and the SS-25 is a new phenomenon, unique to the period since 1979. Since a sharp downturn in United States-Soviet relations also dates from 1979, the worsening of Soviet compliance behavior probably is associated with the changed climate of Soviet-American relations.

Returning to the question of how internal Soviet decision making for compliance actually operates, it is plausible that in a climate of poor United States-Soviet relations and low prospects for further arms limitations, Soviet political leaders would have been more receptive to recommendations from the Soviet military and other arms control skeptics which emphasized the ambiguity of treaty commitments and the strong need to pursue ongoing Soviet programs in the face of a growing threat from the United States. . . .

The United States should adopt a more careful and realistic approach to interpreting Soviet treaty commitments. In particular, the United States

government and the press should more accurately portray to the public the limitations of Soviet arms control commitments, despite the interest a United States administration may have in portraying Soviet understandings as according with those of the United States. This is especially true when the Soviets have made unambiguous statements about their interpretations of treaty provisions which differ from the official United States view. With this knowledge, the United States Congress and public might be less chagrined when the Soviets proceed to do exactly what they have said they would do in pursuing a weapon system or defense program.

A Quiet but Firm Response

Additionally, the United States should respond quietly but firmly to slight Soviet infractions, such as the location of the Pechora radar, so that clear signals about United States understanding of the meaning of treaty limits are conveyed to the USSR. Such an approach might decrease the ability of individuals close to the Soviet leadership to argue, as they may have done in the case of Abalakova, that the United States response would be insignificant.

In retrospect, the United States should have made every effort to use the occasion of the Pechora deployment to raise the issue of the siting and function of LPHARs to obtain an agreement with the Soviets more precise than that embodied in the ABM treaty. The United States should similarly use the Krasnoyarsk issue in the SCC or in negotiations outside the SCC to engage the Soviets in a process of defining specific criteria for distinguishing permitted LPHARs from those that are prohibited. Lacking such detailed criteria, the problem of separating allowed from disallowed radar is going to grow worse in the coming years, as both countries experiment with components that could be employed in strategic defense systems.

Resolution of the big three current compliance problems, as well as prevention of further occurrences, depends upon the future climate of United States-Soviet political relations. There are some encouraging signs on this front emerging from the USSR. A change in approach in Soviet compliance policy has been visible since the beginning of the Gorbachev regime in March 1985. There have been no new significant instances of Soviet noncompliance since that time. Several actions have been taken by the Soviets to resolve compliance disputes of lesser significance, including reaching an SCC agreement restricting concurrent testing of ABM and SAM [surface-to-air missile] components, and possible removal of prohibited SS-16 missiles from the Plesetsk test site. The Gorbachev leadership has essentially admitted that the USSR has a compliance problem in the Krasnoyarsk radar; they have asked to discuss the status of the radar with the United States together with the status of two United States LPHARs currently undergoing modernization in Greenland and the United Kingdom.

Gorbachev's Impact

Within the first year of his regime, Gorbachev effected a thorough house-cleaning among the top levels of the Soviet military that remained from the Brezhnev, Andropov, and Chernenko periods. Was the Soviet tendency toward noncompliance part of the policies pursued by the top military leadership that Gorbachev sought to redirect through his personnel changes? We can speculate that there might have been a Soviet Compliancegate at the onset of the Gorbachev administration, at which point the slide toward noncompliance was halted.

More importantly, Gorbachev's strong desire for a new arms control agreement and his expectations for improved relations with the United States—key ingredients in the Soviet leader's scheme to accomplish bold domestic economic reforms— increased his need to resolve the outstanding compliance issues with the United States. Clearly, any Soviet movement on the big three compliance issues will be unlikely as long as United States-Soviet relations remain in a confrontational phase. Any change in Soviet behavior is likely to come about as a result of Soviet incentives to be forthcoming in the context of their desire for new agreements and less hostile relations with the United States. . . .

The extent to which compliance has become a sore point in United States-Soviet relations and has raised doubts in the United States about the reliability of the USSR as a negotiating partner, indicates that the two countries will need to pursue settlement of the outstanding compliance problems as a first step toward a new arms accord. Agreement on the issues that have emerged from the compliance dispute— more precise limits on LPHARs, more airtight limits on encryption and on ICBM modernization, and many other possible measures to strengthen the existing arms control regime—would provide an appropriate and convenient series of successful accords for another Reagan-Gorbachev summit meeting. Formal United States and Soviet recommitment to the ABM treaty and other accords, together with agreement on updating and rendering much more precise the provisions that have come under pressure from the two countries' military programs, appear increasingly possible today because of Soviet incentives for a new arms control agreement. Such a set of agreed interpretations of past treaties could smooth the road toward new accords.

Gloria Duffy is head of Global Outlook, an organization that does public interest research and consulting on international security issues. She has also worked at the Rand Corporation and the Arms Control Association.

viewpoint 9

Defense Spending Increases US Security

Frank C. Carlucci

[In 1987] Americans celebrated 200 years of constitutional government and the many blessings it has secured. The world has changed a great deal since 1787, and the United States has become a world leader in ways hardly imaginable by our forefathers. America is no longer an island nation with the option of "splendid isolationism." To prosper, we have had to bridge our ocean moat. Our economic and political interests have become global. To defend our values and our interests we have formed alliances, and maintained forces overseas.

The Department of Defense (DoD) exists to fufill the national government's first obligation: to secure the nation's survival and independence against hostile powers that threaten our way of life. Our mission is to preserve America's freedom and secure its vital interests, creating an environment that allows our nation to prosper. . . .

Interests and Commitments

U.S. national security interests are derived from broadly shared values (e.g., freedom, human rights, and economic prosperity) that serve to define specific interests and associated geographical concerns (e.g., the territorial integrity of our allies, unencumbered U.S. access to world markets and sources of strategic resources).

America's preeminent national security interest is the survival of the United States as a free and independent nation, with its fundamental values and institutions intact, and its people secure. We also seek to promote the growth of freedom, democratic institutions, and free market economies throughout the world, linked by fair and open international trade. More specifically, we support the security, stability and well-being of our allies and other nations friendly to our interests. We oppose the

expansion of influence, control, or territory by nations hostile to freedom and to other fundamental values shared by America and its allies.

Other key American interests include:
• Unimpeded U.S. access to foreign markets and resources, and to the oceans and space; we also support the same free access for our allies.
• Consistent with other U.S. interests, reductions in the levels of armaments throughout the world.
• The peaceful and favorable resolution of disputes affecting U.S. and allied security.
• The open exchange of ideas and other measures to encourage understanding among the peoples of the world.

The protection of these interests has, over the years, led America to enter into commitments with other nations in the form of international treaties and agreements that secure and protect those interests. Alliances like the North Atlantic Treaty Organization (NATO) and bilateral agreements like those we have entered into with Japan, the Republic of Korea, the Philippines, Thailand, and Australia, serve to defend those common values that we share. By defending ourselves in this collective manner, we not only improve our own security, but we do so at a reduced cost; common security programs benefit all, and the defense burden is shared by all.

The defense program is designed to safeguard America's values, protect our vital interests, and fulfill our international commitments. Any reduction to this program must weigh any anticipated short-term savings with the increased long-term costs we are certain to incur, along with the increased risks to our interests and commitments.

Instruments of Power

America's defense policies are a component of our National Security Strategy, which is our overall plan for using all instruments of national power to ensure U.S. security. These instruments are interwoven with

Frank Carlucci, "Report of the Secretary of Defense Frank C. Carlucci to the Congress on the Amended FY 1988/FY 1989 Biennial Budget," February 18, 1988. Washington, DC: U.S. Government Printing Office.

our military power and include:

• **Political Power:** America's influence with nations around the world has developed over many years. We have proven many times the non-threatening character of our aims. Of course, America's political power depends to a considerable extent on its military power, since the latter underwrites the stability that makes possible our exercise of peaceful, political influence. Moreover, our military strength provides a counterweight to the threats of our adversaries that might otherwise intimidate other nations into rejecting our peaceful political initiatives. In regions where conflict threatens U.S. interests, our diplomatic efforts—drawing on all elements of national power—can be essential to restoring stability and fostering outcomes favorable to us and to our allies and friends. In these instances, America's military posture can and does provide significant support for these efforts. For example, in the Persian Gulf our willingness to employ military power to protect our interests increases our influence with nations in that region. A strong U.S. military posture, backed by domestic political support and resolve to protect our interests, conveys the firmness of our commitments to allies and friends, thereby enhancing deterrence and increasing the incentives for adversaries to negotiate seriously toward outcomes favorable to us.

"Our nation relies on military might for defensive purposes only."

• **Economic Power:** America's economic strength is an enormously important instrument of our national power. Our economy provides the wherewithal for producing and maintaining our military power. In times of crisis or conflict, the economic strength of our industrial base can be mobilized as necessary to expand and sustain our military power. Our economic power not only supports our military strength, but also provides a basis to secure our interests through other means. It is a major source of our leadership and influence among allied and friendly nations, by which we can promote the international economic stability that is conducive to military stability. Also, it provides resources to support U.S. foreign assistance and security assistance programs.

In an ideal world, the vigorous exercise of non-military instruments of national power would be the means of competition in the global arena. Regrettably, some nations still view the use of armed aggression as a legitimate means of advancing their ambitions. In contrast, our nation relies on military might for defensive purposes only. We see the exercise of military force as a last resort, not as a

substitute for other instruments of national power. Moreover, the United States seeks to use all its non-military instruments of national power to reduce the threats driving our military requirements. To achieve adequate security, we would prefer to lower our military requirements rather than add to our military capabilities.

But our goal of reducing our defense effort while enhancing our security is, at best, an uncertain and long-term proposition. Paradoxically, history has shown us that the best chance we have to negotiate reductions in security threats is by showing our resolve to counter them, by force if necessary. A recent example of this paradox is the agreement to eliminate all U.S. and Soviet ground-launched ballistic and cruise missiles with ranges between 500 and 5,500 kilometers. The agreement was reached only after we began deploying intermediate-range nuclear forces (INF). By maintaining an adequate defense posture, we convey to our adversaries that they have nothing to gain by trying to achieve a military advantage over us. An adequate defense posture also preserves our security until the time when our non-military efforts can reduce the threats facing us.

The successful conduct of U.S. national security strategy involves the integration of all instruments of our national power. *But it is our military strength alone that creates a secure environment allowing us to employ the other elements of our national power in attempting to preserve our security through peaceful means.* Thus, our defense posture is designed to underwrite an international order in which peaceful commerce and diplomacy, not military force, guide the fate of nations. . . .

Risk and Affordability

Defense planning is not a precise calculus, and a nation can never be perfectly safe. In any case, the high cost of military forces, combined with our limited resources, usually means that we must accept some degree of risk—the gap between our defense capabilities and our best estimate of defense requirements. Our goal is to keep that risk at a prudent level.

The appropriate level of security risk for a nation must be decided with great care. While we would like to reduce the risks to our security interests to an absolute minimum, we must also recognize that we have entered a period of constrained resources that will see our military force structure shrink and our overall defense capabilities reduced. The result will be significantly greater risks to our ability to achieve our strategic goals. Thus we face difficult choices regarding our defenses. How well we make those choices and how well we manage their implementation will determine, to a great extent, whether or not we will preserve the gains of the past seven years and build upon them to provide for a more secure America.

Economic considerations do have a significant bearing on the resources we devote to defense. Yet we must also remember that the defense efforts of our principal adversary, the Soviet Union, greatly affect the level of resources we require to maintain the degree of risk to our security at a prudent level. . . . Although the United States has in recent years restored a level of investment approximating the Soviet level, the Soviet Union retains most of the equipment, facilities, and designs they acquired by their much greater cumulative investments since 1970. . . .

"To achieve America's defense policy goals and preserve the common defense, we have a sound military strategy that guides the development, acquisition, and deployment of U.S. forces."

To achieve America's defense policy goals and preserve the common defense, we have a sound military strategy that guides the development, acquisition, and deployment of U.S. forces in peace and, if need be, in war. The manpower and materiel programmed to execute this strategy must be sufficient to preserve our security against existing and near-term threats. Our defense budget states what resources are required to carry out our strategy.

Non-defense resources for ensuring the health and welfare of our people are provided by federal, state, and local governments, and the private sector. If resources for these purposes are limited in one sector, another can make up the difference. Only the federal government, however, provides for our national defense. If the federal government does not allocate sufficient resources for our security, or does not provide these resources in a timely manner, no other source will make up the shortfall. Therefore, our collective purpose must be to request and enact a defense budget that protects and preserves our freedom and security in the most efficient manner possible, and at the lowest acceptable level of risk. . . .

Revitalizing Our Defenses

It is important to understand clearly the resources and effort that were required to overcome defense underfunding in the 1970s. Then, economic considerations played the major role in setting the level of defense expenditures, as the federal government tried to cope with what many thought to be long-term runaway inflation. Annual double-digit price increases were driving up the cost of

government, especially in entitlement programs, and fiscal restraint was decreed for all areas of federal spending. Unfortunately, reducing the defense budget became the primary means of lowering federal expenditures and, consequently, real defense spending declined by over 20 percent from 1970 to 1980.

Our failure to provide adequate defense resources in the face of a growing threat left us with a force structure insufficient to execute our military strategy. Our troops' combat readiness deteriorated; critical modernization programs were deferred; and investment in future defense capabilities was seriously curtailed. Low military pay and poor quality of life for service families led to recruiting problems and an exodus of highly trained personnel. Readiness and sustainability suffered as training, spare parts, and ammunition were not adequately financed. Acquisition inefficiencies prevailed as production runs were stretched out, weapon systems' costs sky-rocketed, and new technologies failed to move into production. America's prestige declined, and our allies questioned our ability to meet defense commitments. Unfortunately, these defense reductions did not contribute appreciably to keeping overall federal expenditures in line since non-defense spending kept rising. Furthermore, while America was cutting back on the resources it devoted to defense, the Soviet Union was engaged in an ambitious and sustained program to expand and modernize its military forces. Thus, paring America's defense resources proved not only an ineffective economic policy, but also an unsound national security policy.

In 1981, the American people and the Congress recognized the dangers inherent in our weakened defense posture, and strongly supported President Reagan's plan for rebuilding U.S. defense capabilities. The President's program redressed our immediate defense shortfalls while laying the groundwork for a long-term modernization program. It has taken seven years and nearly $2 trillion to redress the short-sighted budget cuts of the 1970s and restore the defense capability our security requires—a well-manned, well-equipped, balanced structure of strategic, general purpose, and special operations forces capable of deterring aggression across the entire spectrum of conflict, or of securing U.S. objectives should deterrence fail. Our more robust military posture promotes our leadership among allied and friendly nations, and strengthens our hand in dealing with adversaries.

Defense and the Deficit

Over the last few years, domestic economic considerations once again have dictated the level of defense spending, as reducing the federal deficit became the primary objective in setting the level and allocation of federal resources. Congressionally approved levels of defense funding have decreased in real terms for the past three years, for a cumulative reduction of 10 percent over 1985 levels. We are now again witnessing the consequences of defense underfunding through the reemergence of problems in our force readiness and modernization. Just as in the 1970s, we are being forced to delay important programs, reduce training, defer maintenance, and curtail plans to complete stockpiles of ammunition, spare parts, and other essential equipment. These stretchouts are adding greatly to our costs, thereby inviting further reductions and stretchouts, and jeopardizing the fulfillment of our modernization plans to meet future threats. . . .

Total Federal Outlays

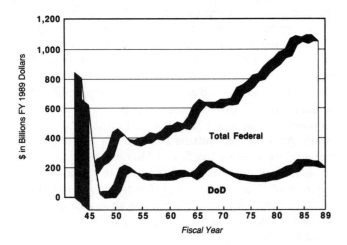

It remains my firm belief that providing sufficient resources for defense is not the principal cause of unrestrained deficit growth; by extension, providing insufficient resources for defense cannot be the principal solution. Recent history proves this point. During the 1960s, for example, when almost 50 percent of the federal budget was devoted to defense, federal deficits were almost nonexistent. Yet in the 1970s, when defense accounted for 25 percent of federal spending, the deficit was on the rise. Today, even though defense claims a slightly higher share of federal resources than it did during the 1970s, the deficit is still higher.

An objective review of federal spending in the first half of the 1980s shows that America's recent defense buildup was not funded at the expense of vital domestic programs. The growth in defense spending from 1981 to 1985 reversed a long-term trend of no growth in defense budgets (see chart). Nevertheless, non-defense spending growth consistently exceeded defense spending growth from 1981 to 1985.

It is also important to underscore that affordability,

in terms of the share of our gross national product (GNP) devoted to defense, is not at issue. In fact, fiscal constraints are forcing a reduction in the share of GNP we allocate to defense, from an average of 6.2 percent from FY [fiscal year] 1982-FY 1987, to only 5.7 percent in FY 1989. This modest level of investment is 25 percent less than the level of GNP we invested in our security during the early 1960s, and less than 1 percent more than the average share of GNP invested in defense during the 1970s. It is also less than half the share of national wealth the Soviets allocate to military spending.

Conclusion

We recognize that the current economic and political environment necessitates reductions to defense budget requests. We have made the necessary adjustments to maintain an overall balance in our defense programs while continuing modernization and development efforts so critical to the success of our deterrent strategy. We cannot, however, ignore the fact that we have been forced to accept a higher level of risk to our security.

Whether or not this higher level of risk proves disastrous depends on our commitment to return to stable and consistent levels of defense funding.

Frank C. Carlucci is the secretary of defense for the Reagan administration.

"The unthinking equation of military hardware with national power ignores the reality of the last forty years."

Defense Spending Does Not Increase US Security

Richard J. Barnet

After fifteen years of slow and erratic economic growth, the domestic costs of the Cold War legacy have become intolerable. For many years, the productive system, the reproductive system (education), and the physical environment, on which all else depends, have been denied needed investments of capital, imagination, and Presidential attention, on the ground that the endless acquisition of weapons and the maintenance of a huge military establishment had a prior claim. The irony is that unexamined national-security policies, designed to protect the people of the United States from foreign coercion and interference, have had the unintended effect of weakening the quality of life in America, constricting the choices of Americans about how to organize and develop their country, and giving non-Americans increasing control over the United States. The overwhelming response to the federal deficit has been to limit expenditures for schools, assistance to needy women and children, environmental protection, preventive health measures, libraries, even police and fire protection, on the theory that in a time of hard choices these have less to do with keeping the United States "strong" than military expenditures. . . .

Economic Damage

By talking about "national security" as primarily a matter of foreign policy and about the management of the economy as primarily a matter of domestic policy, we neatly hide from ourselves the side effects of our strategies for defending the nation and avoid debate on how to strike a balance between foreign obligations and domestic needs. But the connections between the national-security strategy and the mounting insecurity of Americans are becoming too clear to ignore. The American economy is

Richard J. Barnet, "Rethinking National Strategy," *The New Yorker*, March 21, 1988. Reprinted by permission; © 1988 Richard J. Barnet. Originally in *The New Yorker*.

increasingly tied to a world economy over which the United States exercises progressively less control. We are dependent on non-Americans to finance our budget deficit, which we incur in large measure because of military expenditures, and to lure their money we keep interest rates high. The only way to sell otherwise noncompetitive goods has been to sell them at a discount by devaluing the dollar. Now that America has lost its competitive edge in the world economy, there has been a decline in wages for hundreds of thousands of American workers, because they have exchanged high-paying industrial jobs for lower-paying jobs in the service sector. In consequence, many American families are dropping out of the middle class altogether, as dreams of home ownership, or even affordable housing, recede. Having incurred huge obligations payable in dollars, mostly for defense, and having failed to offer attractive enough goods and services for the holders of dollars, the United States finds that its I.O.U.s are being redeemed for pieces of the nation itself. Some of our richest farmland, choice city blocks in metropolitan areas, and major American-based corporations are passing into foreign hands.

The real choices facing the United States in 1988 have almost nothing to do with the arcane matters that provide employment for the Cold War strategists. Far more basic questions of national identity and purpose are at issue. What is the mission of the United States as it enters a new century? To fight Communism? To be the architect and guarantor of a new global order? To build an island of prosperity and freedom in a world of accelerating misery? To force changes in the Soviet Union? The answers to such questions depend not upon knowledge of classified facts but upon wisdom and judgment. How the questions are answered will set the course for the United States in the twenty-first century, now just thirteen years away. Thinking about fundamental choices and goals is never easy, but we make it much harder than it needs to be, because the words we habitually use in talking about national security keep getting in the way.

Since the National Security Act of 1947, the term "national security" has become a talisman. In the name of "national security," telephones are tapped, mail is opened, countries are invaded, American citizens are put under surveillance, Congress is deceived, the Secretary of State—perhaps even the President—is deceived, and, in the Nixon era, high crimes and misdemeanors were committed. But in no statute is there a definition of "national security." It has been enough to say that the United States has enemies, and that the way to deal with enemies is to acquire and from time to time use armed force against them. The only even quasi-official definition of "national security" I have found is in a dictionary that was prepared for the Joint Chiefs of Staff: "a) a military or defense advantage over any foreign nation or group of nations, or b) a favorable foreign relations position, or c) a defense posture capable of successfully resisting hostile or destructive action from within or without, overt or covert."

"The prime objective of modern warfare, the demoralization of the civilian population, has been accomplished in . . . farming communities and city neighborhoods by our own efforts."

This obscure government document merely reflects the conventional understanding in the corridors of power. It would not be much different in the C.I.A. [Central Intelligence Agency], the State Department, or the National Security Council. Security is treated as a scarce commodity. If I have more of it, you must by definition have less of it. The inevitable result is that each player seeks more security by making his potential adversaries more insecure. This is obviously a prescription for permanent war in one form or another. At times, this truism is recognized. Sometimes governments talk about "interdependence" and "common security." But normally the institutions of government and the official rhetoric operate as if the pursuit of security were a contest.

The Nation Is Weakened

In the late nineteen-eighties, we have lost our sense of what "national security" is because we are confused by the meaning of "nation." Nationalism is still the strongest political ideology in the world, but what a "nation" actually is today differs dramatically from what it was in 1789, when George Washington first took the oath of office, and even from what it was in 1948. Any sensible national strategy must begin with some conception of the national interest, whether explicit or not, and this is based on a series of assumptions about what a "nation" is, what it can do, and what costs it may safely incur. Usually, these assumptions are unstated and undebated, because they are hidden in the interstices of language.

A possibly useful definition of "national security" might be something along these lines: "a desired state of physical safety, economic development, and social stability within which there is space for individuals to participate in building an American community free of both foreign coercion and manipulation by government." (Citizens are not secure if the strategy for keeping foreign enemies at bay causes the government to impugn the patriotism of dissenting citizens, and to treat them as enemies to be watched.)

The meaning of "national security" is obscure because the word "nation" presents a false face. As it is habitually used, it is misleading in two respects. First, it suggests that a nation is a distinct entity, which can act on the outside world without feeling boomerang effects, although these are in fact inevitable. One of the consequences of seeing foreign policy as a succession of confrontations involving winners and losers is that the defeats are ignored and the victories—such as the restoration of the Shah to the throne in 1953, the invasion of Grenada, the bombing of Libya—go unexamined; their price tag is neither calculated nor disclosed, nor is the transient nature of the triumph noted. The sinews of national strength—American-owned enterprises employing American workers able to compete in the world market and to maintain what ten generations of Americans have built on this continent—have been weakened by neglect. Much of this sacrifice has been deemed necessary to maintain the war-prevention system we call "deterrence," but the sacrifice has caused some of our inner cities to look as if the war had already taken place. The prime objective of modern warfare, the demoralization of the civilian population, has been accomplished in a substantial number of farming communities and city neighborhoods by our own efforts.

Boundaries Not Enough

The word "nation" is misleading in another respect. It implies that each constituent part of the whole owes its primary loyalty either to the government in Washington or to the American people, who, under the Constitution, are sovereign. But in a commonwealth as divided as the United States has been from the start the system has never worked quite this way. In the War of 1812, a good many Federalists would have preferred to see the British win rather than to see the locus of national power pass to the American West. Today, the question of national loyalty is still more complex. We have assumed, for example, that what is good for the largest American-based corporations and banks is good for the American people, on the theory that their wealth redounds to the strength of the nation. But, increasingly, this is not what has happened. Decisions on the part of banks to make quick profits by foolish loans to Third World nations have helped to create a world debt problem that poses a threat to the economy of the United States. American-based corporations producing goods in Taiwan or Korea are now an important part of the "foreign competition" that has forced a disastrous trade deficit upon the United States. The internationalization of the American economy is not a scandal but a fact of life, a more or less inevitable process in the growth of capitalism and the world economy. The scandal is that neither laws nor official rhetoric recognizes the transformation that has occurred, and the nation grows weaker under the illusion that national boundaries protect us. A rethinking of national-security policy requires looking at the domestic impacts of alternative military and economic policies and understanding the connections better, but the federal government has no effective way of doing this, because the arbitrary separation of foreign and domestic policy is scrupulously maintained.

Power and the Military

All this suggests that "defense" is a much more complex process than the word implies. The seductive image that the word throws up is of a piece of real estate with a fence around it and perhaps a snarling dog or two inside: nations can guard themselves either by putting up a shield (SDI [Strategic Defense Initiative]) or, more usually, by arranging to fight potential intruders on their own turf. But borders are hopelessly permeable. They cannot stop nuclear missiles, drugs, or illegal aliens; as the Iran-Contra affair showed, secrets cannot be kept in, nor can embarrassing information from abroad be kept out.

A more apt analogy is the human body defending itself against disease. The primary strength of a nation comes from within. Its essence is spirit, not hardware. Like people, nations are inescapably vulnerable. How well they ward off outside attack depends on their inner health as much as on their ability to control the surrounding environment. As with the human organism, every defensive response has side effects. Yet we have no institutions within the government to evaluate, or even note, the political, economic, and social side effects of the measures we take for our national defense.

"Power" is another word we use too easily in thinking about security. A major function of the military is to back up diplomatic efforts to advance American interests; armed might is considered a

useful instrument for convincing other nations that American solutions to international problems ought to be taken seriously. On the threshold of the new century, it is very easy to say what power isn't. It is not a new weapon that cannot be used. No addition to a stockpile of sixteen hundred strategic missiles with nuclear warheads increases the power of the acquiring nation, because the nation can do no more with the bigger stockpile either to fight a war or to deter a war than it could do with the smaller. Similarly, the acquisition of military "capabilities" (another tricky word) for fighting implausible "conventional" wars is not easily convertible into power.

"We have no institutions within the government to evaluate . . . the political, economic, and social side effects of the measures we take for our national defense."

The unthinking equation of military hardware with national power ignores the reality of the last forty years. There appears to be a steadily diminishing relationship between the ability to destroy and the ability to achieve political objectives, at least for great powers with much to lose. There have been many occasions and many provocations for the use of military force, since there have been many enemies proclaimed and many acts of aggression committed: the Berlin blockade; Soviet military intervention in Hungary, Czechoslovakia, and Afghanistan; terrorist attacks against Americans; and so on. But the actual commitment of United States military forces to battle has rarely been successful in the past four decades. The last clear-cut victory was General Douglas MacArthur's triumphant landing at Inchon, and that victory was negated two months later by the Chinese entry into the Korean War. (That the bungled operation in Grenada, an island of a hundred thousand people with a barely functioning government and a token army, should have made Americans "stand tall" suggests how victory-starved we have become.) Vietnam, the disastrous rescue mission to Iran, and the failure of America's "low intensity" wars to achieve their stated objectives in El Salvador and Nicaragua suggest that America's security problem may be rooted in misconceptions about the proper missions of the armed forces in today's world. As the ability to destroy has increased, so have the political constraints.

The reason for military failures is not lack of bravery or skill, low pay, or wrong weapons but stubborn realities that greatly inhibit the successful use of force by great powers. Most of the threats that our armed forces have been asked to address with weapons—either by shooting them or by supplying them to others—have been essentially political in nature, and have not responded well to our efforts. The tens of thousands of American lives sacrificed in the attempt to subdue North Vietnam and the twenty billion dollars' worth of sophisticated military hardware supplied to the Shah of Iran to keep an American peace in the Persian Gulf are prime examples of the assigning of impossible tasks to the military. It has become clear since the Vietnam War that the American people will not support large-scale military adventures abroad unless an overwhelming case can be made that national security requires it. This reaction to foreign military intervention, which the Reagan Administration has called "the Vietnam syndrome," reflects not pathology but intuitive realism about the limits of military power in the nuclear age. It is a real constraint on the use of military power as long as the United States remains a democracy.

"The United States is no longer the dynamic actor on the world scene."

The only deployments of military force in recent years which could be said to generate political power took place in situations in which force was introduced as a bargaining chip in a negotiation. The Reagan Administration argues that deployment of the intermediate-range missiles in Europe led to the I.N.F. [intermediate-range nuclear forces] agreement, and that the six hundred and thirty million dollars' worth of covert military aid to the Afghan resistance last year and the hundred and fifty million dollars' worth of aid to the Contras in Nicaragua were the decisive factors in bringing about the possibility of a Soviet pullout from Afghanistan and a liberalization of the Sandinista regime. This argument is not provable, but let us assume that it is true. An important conclusion follows: for a great power, the only effective military option in a world of nuclear weapons is a relatively limited show or use of force, preferably on someone else's territory and with young men from some other country doing the fighting and dying. The Soviet experience in Afghanistan, like the American experience in Vietnam, shows how tightly constrained superpowers are in fighting their own wars—constrained not only by domestic opinion but by the risks of escalation and pressures from other nations. In each of these cases, the superpower abandoned its original war aims despite the abundance of available hardware and troops.

The recent report of the Commission on Integrated Long-Term Strategy, convened by the Department of Defense, recognizes this point and recommends more emphasis on "low-intensity conflict" in the Third World. "Low-intensity conflict" is a euphemism for subsidizing non-Americans like the Contras, the Mujahideen of Afghanistan, or the army of El Salvador to fight what are perceived to be American battles in strategic places. But surely a strategy that bases the safety of the richest nation in the world on the sacrifices of desperately poor people in distant lands is a cynical and corrupting form of dependence, unworthy of that nation.

United States Weakened

Measured by the ability to produce beneficial changes in the international environment, the power of the United States has declined precipitously in the last decade, in part because of changed circumstances, in part because of a failure of nerve. The Reagan Administration, despite its big military buildup, has been unable to bring weight to bear on either of the two prolonged wars of the Middle East—the war between Iran and Iraq, and the struggle among Israel, the Palestinians, and the Arab states. One index of the loss of American power is that the national-security managers have become so torn in their policy goals—at once wishing to engineer peace and to punish nations deemed to be "enemies"—that they cannot make a serious effort to use American power to help settle dangerous conflicts. Indeed, the United States has at times aided both sides in the Iran-Iraq war, apparently in the short-sighted belief that it somehow serves American interests to keep two "enemies" busy.

What sort of power is the United States seeking? Power to do what? What is the connection between the exercise of power abroad and the creation of space for development at home? Until we begin to have a national discussion of these questions, the word "power" will remain empty. Meanwhile, at a time when international agreements and institutions for managing the world economy are urgently needed, the United States is playing a passive role. If the existing arrangements were clearly beneficial to American interests, this passivity would be understandable, but now that the United States is the world's largest debtor nation it has the most to gain from reforms to stabilize monetary arrangements and to establish a new set of agreements and institutions to regulate world trade. Unfortunately, however, the United States is no longer the dynamic actor on the world scene. Increasingly, when it takes action at all, it is simply responding to the initiatives of other nations. . . .

Making Choices

The United States has forgotten the war aims it embraced when the Cold War began. As the Soviet Union appears to be moving toward strengthening its own society by making it less arbitrary and less brutal, some American leaders cannot decide whether this is good news or bad. George Washington warned against "passionate attachments," which let habit and emotion rather than reason decide who is friend and who is foe. If ever there was a time to reach a consensus on what the Cold War is about—whether it is a struggle to contain military aggression, to defeat a "threatening" ideology, or to humble what used to be our only rival for global power—that time is now. Examining old habits of mind and passionate attachments to "friends" and "enemies" would enable the United States to bring commitments and resources into balance. Rethinking the purposes for which we have sought power can help erode our decline in influence and credibility. The United States is still immensely strong, and its possibilities are enormous. The society and the economy are maturing. There is no inexorable path toward decline which Americans are doomed to walk. But choices must be made, and they should be made on the basis of judgments about what has already worked to advance the interests of the American community and what has not.

Richard J. Barnet is a writer with the Institute of Policy Studies in Washington, DC.

"Procurement policy reflects more the interests of suppliers than it does concerns of national security."

Defense Spending Is Wasted

Joan B. Anderson and Dwight R. Lee

The Reagan Administration's attempt to increase U.S. military might has been accompanied by enormous waste, a fact that is obvious to all. Who has not seen the reports of $200 ash trays, $640 toilet seats, and $6,000 coffee pots that the Pentagon has purchased for our defense? Unfortunately, these wildly reported examples of military waste are exceptional only in their lack of subtlety. They represent the tip of the iceberg of waste and inefficiency that is draining the country of billions of dollars of resources while providing little, if any, additional defense.

The natural reaction when confronted with the reality of military waste is to seek the identity of those members of the military-industrial complex who are responsible and to punish them with bad publicity, fines, and possibly imprisonment. In recent times, individuals within the military bureaucracy have been reprimanded for dereliction of duty. More notably, temporary sanctions have been imposed on the two major generals in the military establishment—General Dynamics and General Electric—for taking the offensive against the U.S. taxpayer.

There is justification for such reactions. Military waste is often the result of criminal behavior, and the punishment of this behavior is warranted on grounds of both deterrence and justice. It would be naive, however, to believe that military waste is primarily a function of criminal elements in the military-industrial complex. If fraud in military expenditures were eliminated, there would certainly be some reduction in the misallocation of funds, but the fundamental problem of wasteful military spending would remain.

The technical and secretive nature of the national

defense service provided by the military makes it difficult for the public to exert control over its political agents who formulate defense policy. This difficulty creates a decision-making vacuum that readily is filled by those whose interests are tied directly to military appropriation. Many of the practices that have grown up around defense spending are understood best against the backdrop of special-interest control over national defense decisions—a special-interest control that leads to excessive defense budgets and wasteful allocations within those budgets.

In contrast to most goods, once national defense is provided for one individual in the country, it is provided for all. The fact that one U.S. citizen benefits from military protection in no way reduces the benefits available to other citizens. Goods that have the characteristic of being nonrival in consumption commonly are referred to as public goods. . . .

The political impetus for military spending comes primarily from those who supply national defense, not from the citizens who benefit from it. It is useful to consider in some detail the reasons for this supply-side influence and examine the implication it has for efficiency in military spending.

Special Interests

Decisions on the size of the defense budget and the allocation of moneys within that budget are made under the interacting pressures exerted by four major interest groups: voters/consumers in whose names the expenditures are made; the suppliers—namely defense contractors; the customers, which consist of Pentagon and Defense Department officials and other military leaders; and politicians, acting as middlemen. This interacting pressure is driven by the private interests of these groups and their relative ability to push the growth and allocation of military appropriations in directions

Joan B. Anderson and Dwight R. Lee, "The Political Economy of Military Waste." Reprinted from USA TODAY, May 1987. Copyright 1987 by the Society for the Advancement of Education.

which serve these interests.

While voters/consumers are numerically the largest group as well as the group that pays the bill, their influence is far less than indicated by the standard view. The public votes only indirectly on military programs through its choice of candidates for political office. Each voter also realizes, at least at the subliminal level, that his or her single vote is extremely unlikely to determine the outcome of an election. There is no real incentive, then, for any individual voter to invest the time required to be well-informed. This "rational ignorance" that reduces the voter's ability to direct and monitor political decisions supposedly made on his behalf is especially pronounced in the case of defense spending. There are few, if any, other areas of public decision-making that involve as much technical complexity as decisions on defense. Further aggravating this problem is the shroud of secrecy (some justified, but much surely not) that surrounds political decisions on military expenditures.

"The public's interest in efficiently providing an adequate defense has little influence on defense policy."

In contrast to the weak, diffused, and largely uninformed effect of voters/consumers, defense contractors have strong incentives to exert influence. The Pentagon is the largest single purchaser of goods and services in the economy, with its purchases concentrated in a few industries. Each of these industries has an intense interest in all of the specifics of military spending decisions and in influencing these decisions. They develop the ability to exert influence by maintaining close contacts with the "customers" (the Pentagon), through the use of Political Action Committees (PAC's), and through careful geographical planning in awarding subcontracts. This builds political support at Congressional levels and at the grassroots level, among labor unions and workers.

Defense contractors also gain influence through this easy access to information. Advance information on the types of forthcoming programs not only can give established contractors a head start on securing contracts, but allows them to have input into the policy formation process. This two-way flow of information works to the advantage of the major defense contractors and is enhanced by the "government relations" offices that they maintain in Washington—offices well staffed by former Pentagon and/or Congressional personnel.

Military leaders, along with civilian Pentagon and Department of Defense (DOD) officials, strongly influence the decision-making on military spending. Pentagon and DOD officials are charged with developing military strategy and deciding on military needs. Their decision-making can not help but be influenced by their self-interests—interests that always are attached to the prospects of particular programs and spending proposals. Many Pentagon officials also appear to have personal incentives to accommodate defense contractors who may be former and/or future employers. This so-called "revolving door" has had frequent and extensive use.

The politicians play the role of middlemen in this process. While they are the representatives of the voters, money remains the "mother's milk" of politics. Politicians depend on heavy campaign contributions to survive in political office. Over time, the cost of campaigning, with the extensive use of media coverage, has escalated, making politicians increasingly sensitive to the needs of raising adequate campaign funds. The average voter, as suggested above, has only a marginal role and, hence, a marginal incentive to contribute. As a means of self-preservation, the politician must be sensitive to the desires of large contributors—*i.e.*, the organized interests on the receiving end of political appropriations.

The combination of ineffective consumers and tight cooperative arrangements between the suppliers and buyers of military equipment and the political middlemen who expedite the transactions ensures that the public's interest in efficiently providing an adequate defense has little influence on defense policy. In this regard, national defense presents more of a problem than most goods that the government provides. When the government provides goods that benefit a relatively small and politically organized group of voters/consumers, that constituency will exert political pressure for efficiency. This pressure is weak in the case of national defense because of the broad-based and intangible nature of the benefits provided.

The special interests that make up the defense contractor-Pentagon-Congress coalition possess broad latitude to pursue policies and fund programs that are more effective in promoting their private advantages than in enhancing national security. The result has been an excessively large military budget with a bias toward high-tech, high-profit weaponry instead of a less glamorous, and less profitable, emphasis on conventional weaponry and manpower.

The Military Offensive

This failure in the political process, resulting in excessive and distorted military budgets, is being brought into focus in the current budget debate. An effective push for increased military expenditures beginning in 1979 had increased military spending from $116,300,000,000 in that year to

$265,800,000,000 in 1986. . . . The deficit rose from $40,200,000,000 in 1979 to $202,800,000,000 in 1986. In other words, military expenditures increased by $149,500,000,000, while the deficit increased by $162,600,000,000. Increases in military expenditures have been financed by borrowing (the path of least political resistance) so that, until now, the full costs of these expenditures have been postponed. Politically, it is easy to sell voters on more defense spending when the true costs are obscured.

With the advent of the Gramm-Rudman-Hollings bill, which requires decreases in the deficit, the real choices involved in this type of military build-up are forced. However, if the choices follow the pattern of recent years, the collective wills of the contractor-Pentagon forces will remain much stronger than those of the voters/consumers. For example, a Harris poll in June 1985 found 69% of those questioned preferring to cut defense rather than aid to education. Notwithstanding, in the fiscal year 1986 budget, the share going to defense increased from 26.7% to 27.1%, while the share going to education remained a constant 3.1%. The trend has been to increase defense spending despite prevailing public opinion.

A major inefficiency created by the customer-supplier-political middleman interaction is reflected in the shift in resource allocation within the defense budget. While the share of the military budget going to personnel fell from 36.2% to 26.6% and the share for operations and maintenance fell from 31% to 27.9%, the share going for procurement increased from 17.9% to 29.1% between the fiscal years of 1976 and 1984. (Procurement is defined by the Department of Defense as "the acquisition of weapons, equipment, munitions, spares, and modification of existing equipment.")

"Once major weapon systems begin to be funded, they rarely are stopped, even after they become obsolete."

This shift came in spite of the Joint Chiefs of Staff's call for a substantial build-up in conventional forces, which are said to be "dangerously" thin, given U.S. commitments abroad. Although current policy is aimed at rapid intervention into the world's trouble spots, military budgets emphasize the build-up of heavy, high-cost, limited-quantity weapon systems unsuited for this type of military action.

The reason for this misallocation is that technologically sophisticated weaponry is a very profitable business. It is highly capital-intensive, so labor costs are relatively low. Furthermore, since, under current arrangements, the government supplies a portion of the tools and facilities needed,

capital requirements are relatively low. Risks are also low. Past history of government bailouts makes risk of failure remote. Furthermore, much of the contracting is done on a cost-plus basis, eliminating risks of loss. The result is a U.S. defense budget that is driven by a "technological imperative." It appears that special-interest pressures dictate a military strategy designed to maximize the use of exotic, high-tech weapons, rather than an efficient way to achieve security.

Weapons for Profit

A recent illustration of designing strategy to fit a profitable high-tech system is the Strategic Defense Initiative (SDI), or "Star Wars." The program has the potential of being the first trillion-dollar weapon system. The Council on Economic Priorities has charged that the Pentagon is using its initial contracts to build political support, with 77% of the major contracts awarded to firms in states or districts represented by members of the armed services committees or defense appropriation panels. Defense contractors are calling it the business opportunity of the century, and several of the large contractors—including Lockheed and McDonnell Douglas—have set up special divisions to build political support and secure a share of the contracts.

SDI is the brainchild of physicist Edward Teller, the director of the Lawrence Livermore National Laboratories. Its promises of creating a "protective shield" against nuclear attack and eliminating the need for nuclear weapons are politically appealing. However, many scientists, including Nobel laureate physicist Hans Bethe, argue that it could provide at best a very porous umbrella. The goal of intercepting all incoming missiles is virtually impossible with our current and foreseeable levels of technology. For example, low-flying cruise missiles carrying nuclear warheads could fly under the shield. Bethe also argues that the system could be swamped by multiple warheads and decoys. Since the system works by attacking missiles in their burn phase, the Soviets also could foil the system by designing a missile with a faster burn phase.

[Former] Secretary of Defense Caspar Weinberger, in his *Annual Report to the Congress, Fiscal, 1987*, stated: "Even a thoroughly reliable shield against ballistic missiles would still leave us vulnerable to other modes of delivery, or perhaps to other devices of mass destruction." In other words, a successful SDI, as Bethe suggests, would force the Soviets to change their strategy and their mix of weapons. Is this worth a trillion dollars? Not according to former Secretary of State Dean Rusk, who (in private conversation with one of the authors) has said, "The movement of the arms race into outer space would be politically inflammatory, militarily futile, economically absurd, and aesthetically repulsive. Otherwise it may not be a bad idea." Nonetheless, the system is expanding, breathing new life into the Livermore Laboratories and bringing hope, along with research funds, to defense contractors.

Permanent Funding

Given the supply-determined nature of procurement decisions, once major weapon systems begin to be funded, they rarely are stopped, even after they become obsolete. One such example is the B-1 bomber. Scheduled to come into service in 1986, it is being displaced by a new advanced-technology, radar-evading bomber, the Stealth. Even though the B-1 is expected to be displaced within five years or less and its capabilities could be matched by an upgraded B-52 armed with cruise missiles (both of which also have been funded), it continues to be funded. In fiscal 1985, we spent $5,000,000,000 for procurement of 34 B-1 bombers plus $463,000,000 on additional development. In addition, $458,000,000 was spent to upgrade 90 B-52 bombers. From 1981 to 1988, $28,500,000,000 is expected to be spent on the B-1. It is not only "old soldiers who never die," but also obsolete (and expensive) weapon systems. . . .

This misuse of military appropriations and abuse of U.S. taxpayers are the completely predictable consequences of the nexus of special interest control over military spending. Procurement policy reflects more the interests of suppliers than it does concerns of national security. The waste and inefficiency that have become the hallmark of the military-industrial complex are the expected outcome of the divergence that exists between private interests and the public interest when suppliers are not held accountable to the desires of consumers.

Joan B. Anderson is an economics professor at the University of San Diego. Dwight R. Lee is an economics professor at the University of Georgia, Athens.

"The way the military buys and builds its systems will always be subject to debate, but the worth of the work is unassailable, and the products speak for themselves."

Defense Spending Is Not Wasted

Bernard P. Randolph

Editor's note: The following viewpoint is excerpted from a speech by Bernard Randolph in which he discusses government spending and defense programs.

Like private industry, the military is always striving for a strong, positive public image. Unfortunately, much of what the average citizen learns of military programs comes from the media. When that information is incomplete or sensational—and it often is—folklore is confused with fact. Then, the public consensus that underwrites defense needs is eroded.

Since the current budget trend—toward zero real growth—may force severe cuts into our warfighting muscle in the near term, the Air Force wants an informed public. So I'd like to talk about folklore versus fact, and highlight some truths about defense programs that haven't made the news. As taxpayers, you need to know what's going on.

There are five myths I'd like to dispel: one, that defense is responsible for the deficit; two, that defense programs always overrun their budgets; three, that our warfighting systems don't work right; four, that the military pays too much for spare parts and small tools; and five, that military research and development has no practical use.

Defense and Deficits

I'll start with the budget. You often hear that defense spending caused the national deficit. It didn't, and the Office of the Secretary of Defense recently published a lucid analysis of the numbers to shed light on the subject. In the 1960s, when defense spending averaged about 9 percent of the Gross National Product, deficits averaged 1 percent of the GNP. In the 1970s, defense spending fell to less than 4 1/2 percent of the GNP, but annual deficits grew—reaching above 4 percent of the GNP at one point and averaging over 2 percent.

Since President Reagan's defense rebuilding effort began, the defense budget has averaged only 6 percent of the GNP—much less than in the 1960s and only 1 1/2 percent higher than in the 1970s. The annual deficit has averaged about 5 percent of the GNP. Clearly, defense did not cause the deficit. In the 1960s we invested a large portion of GNP in defense and the deficit remained low. In the 1970s, as the percentage invested in defense fell, our deficit grew. And the trend persists. In fiscal years 1986 and 1987, Congress reduced the defense budget 7 percent in real terms below the FY [fiscal year] '85 level. But the deficit didn't go down. In fact, the annual deficit in 1986 was more than $7 billion higher than in 1985.

Similar analyses prove that cutting defense dollars has not reduced the deficit. In fact, cuts in defense spending in the last three decades have been more than offset by *increases in non-defense* spending. In a speech on that point, [then] Secretary [Caspar] Weinberger used constant 1982 dollars, which eliminates the effects of inflation: While defense spending dropped during the 1970s, non-defense spending almost doubled—from $283 billion 1982 dollars in 1970 to more than $500 billion in 1979. All the while, annual deficits rose. In 1976, the defense budget hit a low of $153.6 billion and the deficit hit a high of $121 billion. Cutting the military budget did not lower deficits, and won't.

Today, non-defense spending accounts for over half of all federal outlays each year. Defense accounts for 28 percent. Yet a public opinion survey [in 1986] found that the average American believes we spend almost twice as much on defense as we actually do. He thinks 46 percent of the budget goes to defense. You have to go back to the '50s to find defense consuming nearly half of the federal budget. Yet the myth persists, and can influence elected officials'

Barnard P. Randloph, in a speech delivered to the Inland Action Military Affairs Dinner at San Bernardino, California, on November 12, 1987.

votes for or against adequate defense funding. As Senator [Everett] Dirksen once observed, "A billion here, a billion there, and pretty soon it adds up to real money."...

Here's another money myth: Many Americans think that modernizing our nuclear deterrent force takes the single biggest chunk of our defense budget. Wrong. Pay and allowances is the biggest line item. The Peacekeeper missile, the world's most powerful deterrent to unleashing the nuclear genie, consumes less than 1 percent of the defense budget.

Cost Overruns Low

Now, one of my favorite topics—cost overruns in defense programs. The Rand Corporation looked into cost growth and found that major DOD [Department of Defense] weapon system programs have lower cost growth than most civil programs—20 percent lower than overruns on congressional office buildings, 50 percent lower than for nuclear power plants, over 500 percent lower than Concorde or the Trans-Atlantic Pipeline development.

That data is more compelling when you consider that military R&D [research and development] demands more risk taking and uncertainty than other government development projects. Take the comparison with congressional office buildings. They have higher cost overruns, but far fewer technical challenges. We know what building materials cost— marble, 2-by-4s, sheetrock—*and* we know we can get them. But what does the Air Force bite off, for instance, with a high-tech project like a spaceplane now under way in my command?

"There are a lot of uninitiated defense experts out there with theories about why big ticket Air Force programs are run the way they are. They don't let facts get in the way of their conclusions."

Before this century's out, the National Aerospace Plane will fly anywhere in the world in 90 minutes, using low earth orbit. It'll reach Mach 25 to get there, and have all the versatility of the shuttle. To build it, we'll need lightweight materials with strength and stiffness never before achieved.

Then there's the propulsion system—a supersonic combustion ramjet. This'll be the first use of air-breathing hydrogen-powered engines. The only way to know if all systems are go will be by flight testing above Mach 8. Predicting the risks associated with these flight tests falls outside our total experience to date. Controlling costs in this environment is where we really earn our money.

The spaceplane's just a snapshot of the challenges in bringing in a truly high-tech program affordably and on time. Yet the defense team is obviously doing a credible job compared with major civil projects. I have yet to see headlines to that effect.

The Weapons Work

More often than not, you just hear the negative angle on weapon developments. That brings me to the third body of folklore I'd like to dispel—that the military's weapon systems don't work well. There are a lot of uninitiated defense experts out there with theories about why big ticket Air Force programs are run the way they are. They don't let facts get in the way of their conclusions. Let me single out two of the Air Force's key programs for a look at the flip side: the B-1 bomber and Peacekeeper.

B-1 bashing in 1987 focused on performance and technical questions. For instance, one news story said the B-1B couldn't fly any higher than the B-17, or about 20,000 feet. Well, the fact is we haven't designed high-flying bombers since the early 1960s. In today's surface-to-air environment, they can't survive where a 747 flies. The B-1B's designed to get down in the weeds and streak under radar. It's supposed to fly very fast at 200 feet, and does. The 20,000-foot ceiling only applies when the airplane is carrying a full load of fuel and weapons. If there were any reason for going higher—like coming home after leaving hostile airspace—it can climb to higher altitudes.

Critics also reported that the defensive avionics suite—that's the system that warns of threats and lets the crew jam or fool the enemy—didn't work. Reports were that the suite worked like a beacon, attracting the enemy and preventing the aircraft from penetrating to the target.

We have, in fact, had a problem with the defensive avionics. The 118 black boxes totaling 5,000 pounds checked out individually, but didn't function well as a system. We analyzed the glitch, and are fixing it. But the fact is, penetration capability involves more than the defensive avionics suite. It's also a function of radar cross section, speed, altitude. The press failed to mention that compared with the B-52 the B-1B is 50 percent faster at lower altitudes, delivers 33 percent more payload and has a greatly reduced radar cross section.

Those capabilities, coupled with the defensive avionics subsystems that already work fine, guarantee that all alert aircraft at Dyess and Ellsworth Air Force Bases are ready to fly, fight and win if they have to. Not to belabor the point, but the news also said that the B-1B has a weight problem. It flat out does not. It's "dry weight"—the weight without fuel or payload—is only 4 percent greater than the A Model's because the Air Force requested it. We asked Rockwell to beef up the plane to handle greater weapon and fuel loads.

You know, on the Fourth of July [1987] the B-1B broke four existing world records and established 14 others for speed, payload and distance. September 17, it broke nine more records and established nine others. Doesn't sound like an overweight, underpowered airplane to me.

Distorted Reports

The B-1B's had its share of technical gremlins for which it's taken *more* than its share of hits. As for Peacekeeper, it's taken hits lately too, but without reason.

You might have all seen the reports. They all claim that, "A major question mark hangs over the missile's capability." The simple fact is, 17 consecutive successful flight tests have proven missile accuracy today is better than we expected to reach in 1990, and already twice that of Minuteman III. There have been *no* flight failures—unprecedented in any missile program.

Reports also insist that it is necessary to take some Peacekeepers out of their silos to check for defective guidance systems. Well, the guidance systems on alert are not defective. They've been checked at two contractor facilities and at the deployment site, and monitored 24 hours a day while on alert. Plus, removal of the guidance system for inspection or repair, which is done routinely, requires no missile dismantling since the guidance drawer is designed to slide out of the missile while still in the silo. Anyone familiar with the system has been aware of this engineering advantage for years.

An independent study committee composed of heavy hitters from industry and the country's finest universities just finished up an intensive examination of the performance of the Peacekeeper inertial measurement unit. The final report of this scientific advisory board came out on November 4 [1987] and confirms that this is one "remarkably accurate" missile. The report has been virtually ignored by many segments of the media.

Horror Stories

You know, errors in fact about the capability of the B-1B and Peacekeeper, two of the three legs of the strategic triad, have far-reaching effects on the national security. They're the foundation of deterrence, and capricious debates over their merits can have a destabilizing effect—just as using them as political footballs can send confusing signals about our national will.

Of course, the B-1B and Peacekeeper are big ticket programs that are likely targets for critics. But some of our worst press has focused on small buys of hand tools and spare parts, which brings up a fourth myth—that the military pays too much for spare parts and hand tools. You've seen the claims of overpricing, but you may not realize that the initial

account of a "horror story" about an allen wrench or coffee pot just gives a shocking claim. The full picture, when investigated, never gets the same coverage.

For instance, did the Department of Defense really pay $110 for a diode that costs a dollar at Radio Shack? Yes, we bought two. That price was then challenged, and a refund received. Later, we bought 122,000 of those diodes for 4 cents each.

Now what about those $7,600 coffee pots? We did pay that price for several replacement pots for the C-5A aircraft. Part of the problem was our own unrealistic specifications for the unit—it had to withstand 40 Gs, produce 12 fresh batches per hour, and be field repairable. We have since fought a war on overspecification. Such coffee pots are now off-the-shelf items for which we pay less than airline companies.

"There's a flip side to the criticism of the United States military and its partners in industry that people just don't hear."

Finally, what about the allen wrench we supposedly paid $9,600 for? We didn't, because the price was challenged and withdrawn.

I could continue, but the point is there's a lot of folklore out there, and public uproar over procurement horror stories isn't just a flash in the pan. It sticks and builds, putting pressure on legislators to act, resulting in what Congressman Jim Courter of New Jersey calls a "hodgepodge of pork barrel amendments, annoying reporting requirements and dizzying funding restrictions [that] cannot fairly be called responsible legislation."

I'm convinced the media and legislators have the best interests of the U.S. at heart when they scrutinize defense programs. But in their zeal for oversight and obsession with problems, they can miss opportunities to support a strong, affordable defense.

Commercial Publications

Let me finish by cracking a final bit of folklore: that military research and development is directed toward blue-sky, gold-plated high technology that offers nothing to the commercial sector or the average American. Well, the fact is, over the past 30 years, military funded research and development has spawned major new industries and greatly improved the American quality of life.

The personal computer, maybe the world's fastest growth industry, began as part of an Air Force research program into micro-miniaturization and the integrated circuit chip in the 1950s. The digital watch and the calculator, plus the timing control circuitry for most home appliances, cars and computers, also evolved from Air Force-sponsored research in the 1960s.

Modern ceramic cookware that doesn't burn, discolor or crack under sudden temperature changes comes from the same material developed for ICBM [intercontinental ballistic missile] nose cones. Graphite composites used in Air Force aircraft and engines are now found in golf clubs, fishing poles, skis, tennis rackets and race car bodies.

The cutting edge of your electric razor; the carbon lining of pipes; lightweight clothing and luggage; sun glasses that darken as the sun gets brighter—even the cloth-like paper of coffee filters—are all "spinoffs" of Air Force research and development.

The Jumbo Jet evolved from the military need for a wide-bodied cargo carrier—not vice versa. TV instant replay, microwave ovens, satellite weather information and laser technology are all spinoffs. I could continue, but you get the picture. There's a flip side to the criticism of the United States military and its partners in industry that people just don't hear. Our efforts secure strength with freedom *and* a better life for all Americans.

Worthwhile Work

The myths versus realities I've given today are just the tip of the iceberg. I think the way the military buys and builds its systems will always be subject to debate, but the worth of the work is unassailable, and the products speak for themselves. We generate combat systems our friends envy and our enemies fear.

Ten countries fly the F-16. Six others are buying them. We've sold 2,500 F-5s to 30 other nations. We've delivered the Airborne Warning and Control System [AWACS] to NATO and Saudi Arabia. The U.K. recently selected AWACS over their own NIMROD, and France chose it over all other airborne surveillance options.

How do our systems fare in the fog of war? In 1982, using F-15s and F-16s, Israel downed 82 Syrian MiGs, with no losses of its own over the Bekka Valley. F-111s gave Qhaddafi just a *taste* of our warfighting capability during the raid on Libya. And no enemy has risked the full retaliation of our warfighting systems. In the big picture, that's the measure that matters.

Bernard P. Randolph is the commander of the Air Force Systems Command.

"Fraud, waste, and excessive profits are the predictable consequence of turning to private corporations for weapons production.

viewpoint **13**

Defense Contractors Are Corrupt

Center for Defense Information

The U.S. entrusts the production of weapons to private corporations. Since World War II a small cadre of large military corporations has emerged which depends for its survival on winning government contracts to build weapons. They aggressively sell their weapons by applying their sizable political and economic influence. As a result we buy too many weapons we don't need and pay too much for them.

Relying on private corporations to make weapons insures endless procurement scandals and cedes to these contractors a substantial role in the determination of the size and composition of U.S. military forces. Contractors not only control how much we pay for weapons but also influence how many weapons we will buy. They have gained a sizable foothold in the formulation of key national security policies.

While building weapons is now big business, the industry bears little resemblance to free enterprise. Under the present system the U.S. government supplies much of the industrial equipment up front, covers most of the cost, assumes most of the risk, and still pays out higher profits than those earned in comparable commercial work. The present system combines the inefficiency of government management with the profit motives of private enterprise to achieve the worst of both worlds. . . .

Big Business

Making weapons is one of the biggest industries in the U.S. About 35,000 businesses receive contracts directly from DoD [Department of Defense] and another estimated 150,000 do subcontracting work for those firms. Most U.S. industrial corporations are involved in military work. The top military contractors—IBM, General Motors, Ford, Boeing,

Lockheed, Rockwell, and General Electric—read like a list of corporate America's Who's Who. The top 10 Fortune 500 industrial corporations are all major weapons or petroleum suppliers to the Pentagon.

The military industry in the U.S. is enormous by any standard. Military sales were nearly $185 Billion in 1985: $166 Billion to the Pentagon, $6 Billion to the Department of Energy, and another $13 Billion in weapons exports. If it were a separate economy, the U.S. military industry would be the 13th largest in the world, just ahead of the entire economy of East Germany. Private corporations employ about 3.3 million workers for military contracts, as many as the entire population of Nicaragua.

The military industry is divided between the very large firms and everyone else. The top 10 firms account for one-third of the Pentagon's sales and the top 25 contractors account for 50% of all Pentagon contracts. Only the top companies can win contracts to make major weapons systems. The rest of the companies make less expensive weapons or subcontract for the big firms. The largest firms have become entrenched in their dominant position.

Many of the biggest firms are heavily dependent on the Pentagon for their survival. Six major aerospace firms are dependent on the government for at least 60% of their sales. The Lockheed Corporation, for example, used to be in the commercial aircraft business. Since the late 1960s it has essentially become an appendage of DoD, with 88% of its sales to the U.S. government. On the whole, the government buys nearly 50% of the sales of the U.S. aerospace industry. *These corporations have a vital interest in the continued sale of arms and bring all of their enormous economic and political leverage to bear to promote military and congressional support for their weapons programs.*

Top military officials are now confidently predicting that new government regulations have ended the waste and mismanagement scandals of

The Center for Defense Information, *"No Business Like War Business," The Defense Monitor*, Volume XVI, Number 3, 1987.

recent years. Such confidence ignores the fundamentally self-centered orientation of private companies. Corporations are influenced primarily by a variety of economic factors, such as interest rates, inflation, and sales. Government regulations and concerns for national security are not primary motivating factors, even for those corporations dependent entirely on the government.

The primary goal of private corporations is to maximize the return on stockholders' investment. There is no reason to believe that this goal coincides with the economic or national security interests of the government. To protect its interests, the government has tried to control contractors through an elaborate, unbelievably complex set of rules and regulations. These rules, no matter how much they are fine tuned, cannot permanently align the contractors' interests with the government's, nor completely control the behavior of contractors. The resulting fraud, waste, and excessive profits are the predictable consequence of turning to private corporations for weapons production.

Involving private companies in weapons production means that weapons become a product to sell, like toasters or washing machines. Military contractors, like other private companies, spend millions to peddle their wares. Contractors attempt to influence the decisions to buy weapons by a variety of means, such as advertising, hiring former DoD employees, making political contributions, and maintaining expensive lobbying efforts in Washington, D.C.

Four contractors alone spent a total of $31 million on advertising in 1985. During a ten year period (1976-85), TRW, a major military contractor, spent at least $100 million on advertising, while McDonnell Douglas spent $50 million. McDonnell Douglas has increased its ad budget six-fold since 1976.

Advertisements are used to sell specific weapons as well as to polish the company's image. For instance McDonnell Douglas spent $184,000 on ads in *The Washington Post* in 1985, including three half-page spreads on the C-17, a new cargo plane it is trying to sell to the Air Force. Lockheed, McDonnell Douglas' competitor, countered with two full-page ads costing $66,000 extolling the virtues of its C-5B cargo plane. Ultimately these expenses must be recovered in military contracts paid for by the taxpayer.

Political Clout

Contractor hiring of former military officials increased nearly 500% between 1975 and 1985. Only top military officials who take jobs with contractors are required to report their new jobs to DoD. Contractors hired 3,842 former top DoD officials in 1985, up from 3,310 in 1984. The interchange of people between DoD and contractors fosters common perspectives and shared interests and gives

DoD officials incentives to befriend contractors. Decisions on weapons contracts may be based in part on future personal employment prospects and rewards rather than military considerations.

The close economic relationship between the government and contractors is reinforced by the political clout of major contractors. Major military contractors actively contribute to political campaigns in order to sell more weapons. Twenty top military contractors poured $3.6 million into congressional and presidential campaigns in 1984, double their 1980 level.

In addition, 53 Political Action Committees (PACs) sponsored by military contractors gave $7.2 million to national campaigns in 1984. Moreover, contractors gave $387,613 in honoraria in 1985—up 50% from 1984—to politicians for military plant tours, speeches, and appearances. These lobbying and advertising expenses by contractors are later recouped in new contracts to build more weapons.

Profit Motive

The profit motive in a competitive industry gives firms an incentive to reduce the price of a product in order to maximize sales. In the absence of effective competition in the military industry, the profit motive gives companies the incentive to sell as many weapons at as *high* a price as possible.

Although the U.S. has long entrusted its weapons production to private firms, it has been an anxious trust. Unwilling to leave firms completely unregulated, there have been a variety of regulations in the past limiting the profitability of military contracting. Recently, however, military contractors have gotten things their way and most of these restrictions have been lifted, leaving DoD without any legislation guarding against excessive profits for the first time since 1934.

"Involving private companies in weapons production means that weapons become a product to sell, like toasters or washing machines."

The Vinson-Trammel Act was enacted in 1934 in order to put a ceiling on profits but it was never rigorously enforced. Peacetime application of the Vinson-Trammel Act was abolished in 1981 and its wartime application weakened. The Renegotiation Board, started in 1951, was designed to limit the level of profits by requiring renegotiation of contracts that were excessively profitable. The Board expired in 1979.

Since the massive increase in military spending started in 1979 it has been the best of times for military contractors. From 1980 to 1985 annual military sales bounded from $84 Billion to $163 Billion. During that period the industry received a total of $764 Billion in orders from the Pentagon. In the torrent of spending the industry was awash in cash and scandal, registering record profits closely paralleled by record numbers of cases of contractor fraud and waste. *Fifty-nine of the top 100 contractors were under investigation by 1986 and the number of contractor suspensions and debarments doubled from 1984 to 1986.*

Profits Soar

Profits in the military industry are significantly higher than profits for comparable commercial business. This conclusion emerges from two 1985 profit studies by the Department of Defense. As a percent of sales, military work during the past 10 years has been roughly as profitable as comparable commercial work. But when compared with the investment made by military firms, weapons work has been almost twice as profitable as comparable commercial business.

The most comprehensive study of military profits in ten years, the Defense Financial and Investment Review (DEFAIR), was undertaken by DoD in 1985. The findings of DEFAIR are as disturbing and startling as they were misinterpreted and underreported. DEFAIR found that from 1980-1983, military work was more than twice as profitable as comparable commercial work. As a percent of assets, military profits from 1970 to 1983 were 20.5% while commercial profits trailed at 13.3%.

Similar conclusions were drawn, using a markedly different method, by a Navy study which analyzed 22 top military contractors. The study compared the returns on commercial versus military work for 22 major contractors. The study found during 1977-1984 the return on assets for government work averaged 24%, while commercial work averaged 12%. Navy Secretary John Lehman, summarizing the study, remarked, *"Defense contractors are getting fair profits as a percentage of sales, but measured as a return on assets they are grossly out of line."*

DoD periodically makes minor adjustments in its complicated formula to control profit levels on military contracts. Often such minor adjustments have unintended results because the profit levels of private firms also depend on more powerful forces in the economy, such as interest rates and inflation.

A particularly shocking illustration of this was a change in the profit policy in 1980 which precipitated an unintentional $500 million to $1 Billion in excessive profits each year through 1986. DoD has issued a new profit policy that will take effect in 1987, which it claims will stop these excessive profits and encourage greater contractor investment. Had the 1980 adjustment produced a squeeze on profits instead of a windfall, one may be sure that it would not have required six years for the contractors to obtain relief through political channels.

Because it is difficult to predict the exact effect of any new profit policy, there is no guarantee that such errors will not be repeated. As long as private industry is involved, the risk of overpayment, underinvestment, or other problems will exist. A biennial profit review is urgently needed to avoid such overpayments in the future. Only with such a biennial review will DoD be able to track more precisely the effect of its profit policy and be able to make regular adjustments in profit controls to insure that taxpayers are paying fair, not excessive, profits.

Money in the Bank

Profit levels are merely accounting measures, figures on a page, which are subject to manipulation. They do not as such represent money in anyone's pocket. The primary way profits become cash for individuals is through salaries and bonuses, stock appreciation, and stock dividends. In terms of cash for individuals, military contracting has been substantially more lucrative than comparable commercial work.

"As long as private industry is involved, the risk of overpayment, underinvestment, or other problems will exist."

• Salaries and Bonuses: The GAO [General Accounting Office] found in a 1984 study that executive pay at a dozen large military aerospace companies averaged 42% more than the average salaries and bonuses in comparable commercial companies. The average executive salary at aerospace firms was $285,494 in 1982. The lowest annual pay for a Chief Executive Officer among the top ten military contractors was $500,000 in 1983. For instance, David Lewis, CEO of General Dynamics, in 1983 received $487,000 in salary and an additional $898,000 in various bonuses.

• Stocks: The second way profits find their way into people's pockets is through stocks. If a corporation retains its profits, the value of the firm and hence the value of the firm's stock increases. Stockholders' shares become more valuable. The corporation can also distribute the profits through dividends. The best indication of profitability of a firm is its "market return": the sum of a stock appreciation and dividends.

By this measure military work is again substantially more profitable than commercial work. Military stocks averaged a 3.3% higher annual market return than for commercial stocks from 1975 to 1985. In addition, the biggest firms, those which do more than one-half their work with the government, averaged a whopping 25% average annual market rate of return for the past 10 years, 60% higher than for commercial stocks.

Less Risk, More Profits

Profits are often characterized as a reward for a business undertaking risk. Most business endeavors entail risk of failure and subsequent loss of investment. Profits are a necessary incentive for businesses to make a risky investment. No business would risk money if there were no chance for gain. A particular industry's profit level can only be put in context when its risk level is evaluated. The higher the risk the greater the potential profit must be for the business.

In military contracting this logic is turned on its head—*large military contractors get high profits for low risk business.* Risk is impossible to gauge precisely but can be roughly measured in a variety of ways. By many of these measures, military contracting appears less risky than comparable commercial business.

"These contractors . . . lobby for greater weapons spending, wrapping their pleas in the drapery of patriotism and the Soviet 'threat.'"

Contracts for a weapon system fluctuate from year to year with the shifts in congressional moods. Yet these fluctuations appear less volatile than similar ups and downs in nonmilitary commercial ventures. DEFAIR found that military contracting was significantly less risky than comparable commercial business, in most cases by a factor of two. Navy Secretary Lehman said, "Once production has been approved, the Department of the Navy has historically provided a more stable market to its suppliers than found in the commercial free market."

Even if appropriations for a weapon fluctuate somewhat from year to year, a contractor can be nearly certain that some funds will flow each year. Moreover, if fewer weapons are ordered than anticipated, contractors merely charge more per weapon to cover their costs. The few instances when major corporations have lost money on a military contract were caused by gross mismanagement, not fluctuations in military spending.

A Protected Industry

Cancellations of weapons systems have been extremely rare. Commercial projects, by contrast, can flop at any time at the whim of the market. The road to commercial success is littered with corpses of failed products. In contrast, of the hundreds of

major weapons programs in the last 30 years, only 32 have been canceled. Only five programs valued over $1 Billion were terminated during that period. Former Chief of Naval Operations Admiral James Watkins said, "We are one of the most reliable, prompt-paying customers the world has ever seen."

A final safety net beneath military contractors is the assurance that in a bind, Uncle Sam will not let them go down the drain. If a major military contractor is ever in serious financial trouble, the government simply gives it enough money to keep it afloat. Few commercial firms have any such guarantee. Workers in military plants assume the most risk, facing layoffs if a contract is canceled or cut back but the company and its executives are spared. "Either overtly or subconsciously, we don't let them go out of business because of concern about the defense industrial base," said Major General Bernard Weiss, Chief of Air Force Contracting Policy.

By most measures military contracting for major corporations appears less risky than comparable commercial business. Such low risk, when coupled with a history of relatively high arms profits, characterize an industry which differs dramatically from commercial industry. DoD may consider it necessary to prevent key contractors from going bankrupt and shelter them from the risks of commercial business, but in exchange for reduced risk, contractors should settle for substantially lower profit levels.

The military industry has evolved into a cumbersome mixture of the public and private sectors. *Since weapons production has become a business, a group of military corporations has emerged that live a privileged life unknown to commercial firms.* These contractors retain all the rights of private corporations and lobby for greater weapons spending, wrapping their pleas in the drapery of patriotism and the Soviet "threat."

A prerequisite for a more sensible system would be the elimination of contractor influence over the purchase of weapons through stiff regulations on contractor lobbying and advertising. In addition, a separation of government and private resources—into either entirely public or all private firms—is required for the industry to operate efficiently and responsibly. . . .

The U.S. must move beyond the rhetorical debate of "free enterprise" vs. "nationalization" and recognize the special needs and requirements of the weapons industry. The government is irrevocably involved in weapons production. All that remains is to insure that its involvement is productive: that we buy only the weapons we need and pay a fair price for them.

The Center for Defense Information is a Washington-based research organization that evaluates US defense policies and their impact on American society.

"[Defense contractors] are not realizing excess profits, and they accept more government-imposed abuse than most commercial producers would ever tolerate from a customer."

Defense Contractors Are Not Corrupt

William B. Scott

Government acquisition policies imposed by Congress and the Defense Dept. over the last 10 years are severely hampering the ability of U.S. aerospace companies to maintain weapon system technological superiority, and will force a number of defense contractors out of business within the next 5-8 years.

The often adversarial environment existing between government procurement agencies and some contractors is not improving, and will probably continue to worsen, as long as government leaders fail to understand and acknowledge that industry must abide by the laws of good business or get out of business altogether. The restrictive, high-risk, low-profit-potential policies imposed on contractors are effectively reducing the incentive to innovate and even to participate in defense contracting.

While industry tightens its belt and tries to survive in this environment, procurements that emphasize short-term savings continue to roll out of Washington, effectively eroding the long-term U.S. industrial base that has ensured a strong defense posture in the past. And there are few signs that things will get better soon.

The reasons behind today's government acquisition situation are very complex, defy simple solutions, and cause numerous heated debates. Some industry and government/military leaders have discussed these causes in some detail through professional associations, letters, speeches and trade publications, but have failed to grab the attention of Congress, the American public and the general media. Despite attempts to articulate how the industry got into the current procurement difficulty, and what must be done to improve the situation, there is still a strong general perception that defense contractors are routinely raking in excessive profits. The image of the stereotype "lying-cheating-stealing contractor" is shared by much of the general public and many in government.

"Smart" Competition

This image frustrates many executives who have campaigned for years against wasteful government procurement practices that result in expensive weapon systems characterized by poor reliability. Edward R. Elko, recently retired president of Aerojet Ordnance Co., has been a strong proponent of changing ordnance acquisitions to a system approach, giving one firm overall responsibility for ensuring high system reliability at lower cost. He successfully pushed for "smart" competition in the ordnance business, aimed at holding contractors accountable for quality through realistic warranty clauses. And he resents industry's tarnished image.

"[The Administration] never stepped in to defend industry against the allegations of the media," Elko said. "[They] allowed the media to create an image that we are all crooks. And that is wrong. [Defense secretary Caspar] Weinberger had an obligation to put it in the proper perspective and he didn't do it."

An *Aviation Week & Space Technology* survey of contractors in southern California revealed a profile of defense contractors very different from that traditionally presented. Firms committed to the aerospace industry are having to operate with impediments that range from general Defense Dept. procurement policies and procedures to day-to-day frustrations with government inspectors on the shop floor. They are not realizing excess profits, and they accept more government-imposed abuse than most commercial producers would ever tolerate from a customer.

Many contractors are in the unenviable position, however, of only having one customer—the U.S. government. For this reason, many are reluctant to

William B. Scott, "Defense Acquisition Policies Hinder Contractor's Ability to Innovate," *Aviation Week & Space Technology,* March 14, 1988. Courtesy AVIATION WEEK & SPACE TECHNOLOGY. Copyright McGraw-Hill, Inc., 1988. All rights reserved.

speak out in protest, fearing the backlash of government auditors, inspectors and the contracting officer. Ironically, if the situation were reversed—if contractors could somehow subject the government (and taxpayers) to similar practices—the public outcry would be so loud that Congress and the Administration would be forced to take immediate corrective action.

Unfair Government Practices

ITT Gilfillan executives identified some of the thorny policy/procedural issues facing defense contractors today and for the foreseeable future. Backed by other industry representatives, these included:

• Fixed-price development contracts, which place undue risk on contractors. However, strong aerospace and financial industry warnings have prompted recent statements from military leaders indicating that this policy may be changing.

• Risk/cost sharing and multiple-sourcing requirements. A policy of separating development and production awards, while requiring that contractors commit large amounts of company funds to development, significantly increases corporate risk. At the same time, though, the contractor has no legal means to recoup any development-phase investment during production unless he wins that competition. Further, the loser in a competition often emerges as a risk-free second source for the winner's technology. With no assurance of a return on investment, many firms are opting to not participate in the acquisition process at all under these rules.

• Long-term product warranties. Although there is ample justification for requiring contractors to back their products after they are fielded, companies must carefully negotiate warranty clauses, and either assume high risks associated with newer technologies, or opt for proven, safer designs to reduce the risk of failure.

• Revised guidelines for calculating allowable profits. The government has changed the algorithms for determining what profit levels will be allowed on a given program. Aimed at eliminating excessive profits, the new rules make it much more difficult to make a profit at all.

• Intrinsic value of spares. Early in a program, a spare part might be delivered for $100 under an approved contract. Today's regulations, however, permit the government to legally come back a year later and claim the spare part is not worth $100 and demand a rebate for all parts delivered to date. This has actually happened, according to industry executives.

• Smaller progress payments. As a government cost-saving measure—consistent with the concept of greater industry risk/cost sharing—progress payments for ongoing work has been reduced from about 80% to 75%, improving government cash flow at the expense of the contractor's.

• Requiring contractors to pay for tooling and test equipment. This policy clearly ignores the past congressional and Defense Dept. record for periodically canceling programs at any stage, leaving contractors with perhaps millions tied up in tooling or equipment and no possibility of recouping the cost. This situation may be changing, however.

• More pressure to innovate while simultaneously stipulating that data rights for new designs go to the Defense Dept.

• Increased possibility of facing criminal prosecution over contract deficiencies. In the past, the government acknowledged that complex procurements would produce some honest mistakes. When discovered by either contractor or customer auditors, the contractor paid the government what it owed and changes were made to avoid future occurrences. Now there is a tendency for any mistake to automatically be considered as outright fraud, and the penalties are much more severe. Even honest mistakes are treated with criminal overtones.

An Old Problem

In an aerospace industry address, Thomas V. Jones, chairman and CEO [chief executive officer] for Northrop, also enumerated many of these same points and noted that "there is no doubt as to whether these measures, individually or collectively, provide short-term financial benefits to the government. They certainly do. But I would suggest that the short-term benefits are accompanied by dangerous risks for the long-term, to our defense posture, our national security, and the industrial base upon which that security depends."

"There is a tendency for any mistake to automatically be considered as outright fraud."

Some executives consider the present environment "just part of the cycle we've seen before." There is no question that today's policies have historical roots. One aircraft company manager referenced a February 1987, *Wings* magazine article describing the U.S. acquisition environment that existed between about 1920 and 1933, when the government awarded separate contracts for aircraft development and production, requiring competitive awards for each phase. Designs were owned by the contracting agency, which was free to award a production contract to any company that could produce aircraft at the lowest cost.

This effectively killed industry's incentive to innovate, since good ideas were turned over to someone else to produce at a higher profit level than the development phase provided. The emphasis on competition and lowest bid forced contractors to minimize risk and cost by sticking to orthodox designs and outmoded manufacturing methods. This approach made no sense to astute businessmen who elected to focus their innovative talents and corporate resources on commercial endeavors—such as building airline transports—leaving the more mundane, low-profit military business to Curtiss, Martin and LWF Engineering.

Near Bankruptcy

While there is a tendency for defense contractors to see today's acquisition environment as cyclical, some leaders foresee a potentially different outcome this time unless Congress and the Defense Dept. take positive steps now.

"Today's systems are more complex, more expensive to build, and take longer," according to Tom Steulpnagel, a former president of Hughes Helicopter Co. during the period when the AH-64 Apache was developed. Policies of the 1965-75 period, "when Total Package Procurement was going on, pushed a lot of companies into bankruptcy and forced some, like Lockheed, practically out of business," he said. "And it's going to happen again. The [Bell/Boeing V-22] tilt-rotor program is a $2.8 billion package. If Bell misses it by 10%, they've used up the net worth of Bell Helicopters."

Although the advanced tactical fighter (ATF) program is the most visible and frequently discussed example of the government's risk-sharing policy—having industry assume a substantial part of a program's up-front costs—others keep emerging. This steady trend indicates that few changes are in the offing at the contracting officer level, despite recent encouraging statements by Defense Dept. acquisition czar Robert B. Costello.

Jacques Naviaux, director of marketing for Hughes' Radar Systems Group, observed, "From ATF all the way down, the government's been saying, 'Well, how much are you going to invest?' One [solicitation said], 'If you want to be a second source on this [program], you have to do somewhere around $20 million of work free.' That gives you the right to just compete."

Restructuring Procurement

Government pre-contract solicitations continue to be released that reflect strong adversarial attitudes at the contracting officer working level, as well as a poor understanding of basic business principles. Several RFPs [request for proposal] in recent years have required that an officer of the corporation sign a statement assuring, under penalty of perjury, that the company's proposal was fully compliant with all the specifications and requirements detailed in the RFP. Companies submitting bids must agree to this stipulation or be considered noncompliant.

However, specifications called out in an RFP are often conflicting and inconsistent. Many corporate executives reluctantly sign the statement, knowing that a court challenge would highlight the inconsistencies in these specifications, but "on the other hand, who wants to go into court on a felony charge?" Naviaux said.

Another recent RFP suggests that the government's preoccupation with competition for the sake of competition is sometimes only lip service. An Army draft RFP for a "Large Scale Winged Target" simulating a Soviet MiG-27 requires bidders to develop a full-size composite, jet-powered drone aircraft. The development phase will cover the airframe, an autopilot and complete remote control system, enabling its operation as an unmanned target. The same contract will cover production of a "variable" number of targets, as well as provision of flight services on a government-owned range.

A team of small companies has proposed restructuring the solicitation to reduce costs by splitting the procurement into separate contracts—development and test, production and flight services. Headed by Aerotest, a small company based in Newport Beach, Calif., the team maintains that the present RFP is structured to eliminate all but about three large companies. If reshaped, the procurement could realize considerable cost savings by opening each phase to more companies or teams, according to W. R. Laidlaw, Aerotest founder and chairman.

Composed of three small firms—Aerotest; Scaled Composites, a Beech subsidiary headed by Burt Rutan; and I.C. Sim, Inc., which specializes in digital flight control systems and specialized computer simulations—the team contends that the RFP effectively excludes it from bidding on the program by requiring everything in one package.

"Today's acquisition environment can make it very difficult to design, develop, test, produce and deliver a product, thanks to more government regulations and procedures."

"This team is probably better qualified to do the development than any other company—large or small—in the country. And we can't even bid on it, because [under the present RFP rules] we aren't qualified," Laidlaw said. "We're not interested in producing [the drone] or doing range operations. We only want to design, develop and test it. The Army can find plenty of companies who want to produce it and would do a great job. And there's a half dozen small companies that do an excellent job in flight services. But [the Army] is looking for one big guy to do the whole job, denying the government access to the initiative and talent existing in small companies."

Once a company is awarded a contract, today's acquisition environment can make it very difficult to design, develop, test, produce and deliver a product, thanks to more government regulations and procedures. These concern everything from material storage, inspection and tracking to incredibly detailed documentation of labor man-hours—down to the tenth of an hour—expended on each project.

On-Site Inspectors

Perhaps the biggest surprise for a new defense contractor is the pervasive power of on-site inspectors from the Defense Contract Administration Services (DCAS). Charged with ensuring the government receives a product that meets contract specifications, DCAS representatives typically have offices at the contractor's site, periodically inspecting the progress of a production item as well as compliance with procedures. And, some contractors say, they're everywhere. "There are probably more DCAS auditors today than there are FBI agents," one executive said.

"Industry must play by the rules of good business or they can't afford to play at all."

As a Defense Dept. agency, DCAS inspectors are independent of the military service program manager, and tend to be "more abstract. They look at their purpose as having to find something, whether there's really [a valid problem] or not," according to one manager. Documented cases of DCAS representatives stretching their charter include:

• A DCAS letter to a contractor invoking a contract clause governing retention of material removed from government equipment. The letter stipulated that the contractor retain used spark plugs and motor oil removed from government ground power generators. This required packaging used oil and discarded plugs in EPA-approved containers, tagging them, tracking them via computer and storing them in approved warehouses. Toxic substances such as used oil had to have their containers inspected periodically and their condition recorded. The cost of this activity over the life of an aircraft modification program was substantial—and came out of company profits.

• Contractor production workers periodically waiting for a DCAS inspector to arrive before proceeding with the next step of a process. On one program, the contractor was attempting to meet a

hard delivery date for its first modified aircraft. Over a seven-day period, DCAS inspectors kept a 5-10-person crew tied up with repeated inspections that prevented replacement of numerous fuselage panels. Once, invoking the letter of inspection protocol, the DCAS inspector took about 20 min. to check a small compartment with his flashlight and mirror, ultimately rejecting it for "dirt and grease in one corner." The aircraft had rested in the desert at Davis-Monthan AFB, Ariz., for several years, and, after modification, was expected to survive a maximum of 50 flight hours. The cost-effectiveness of the DCAS approach was highlighted when this particular drone aircraft was shot down on its first unmanned flight shortly after delivery to Tyndall AFB, Fla.

• A major aircraft production program, plagued by frequent delays, finally opened a separate charge number labeled "Time awaiting DCAS inspector." At one time, this number accounted for about 12% of shop floor personnel time.

Less Confrontation

Despite the gloomy outlook for defense contractors squeezed between a shrinking budget and aggressive contracting officers, there are signs of positive changes in the next year, including:

• An end to the flurry of legislation that produced many of the unrealistic acquisition policies over the last few years.

• A shift to a more positive, less confrontational environment. Some aerospace industry executives sense that Defense Secretary Frank C. Carlucci recognizes "contractors don't have unlimited funds, and that fixed-price development contracts don't make much sense," Hughes' Naviaux said.

• In some quarters of the government, less rigid interpretations of procurement policies and a greater willingness to listen to industry's concerns. "Recently, I've been able to discuss a lot of these issues with my customers," Charles R. Sebastian, newly appointed president of Aerojet Ordnance Co., said. "When I've shown them—in dollars and cents—why it's in the best interest of the government and the contractor, [the contracting officers] have listened and changed the procurement criteria. Agencies are starting to respond to the alarm that industry must play by the rules of good business or they can't afford to play at all."

William B. Scott is the senior engineering editor at the Los Angeles bureau of Aviation Week & Space Technology.

"SDI already is paying big dividends in research and technology advances."

SDI Benefits the Economy

Grant Loebs

For the future, the promise of the Strategic Defense Initiative (SDI) is that the United States will be protected from nuclear holocaust. For now, SDI already is paying big dividends in research and technology advances. On the near horizon are spinoffs for the space program, civilian industry, and medicine. These include gains in sensors and radars, new ceramics and metal alloys, computer chips with increased capacities, and lasers of increased power and accuracy. Business Communications Company, a market research firm, estimates that the commercial marketing of SDI spinoffs by the private sector of the economy will total between $5 trillion and $20 trillion.

Expectations of bountiful SDI dividends are realistic if the U.S. space program is a precedent. On May 25, 1961, President John Kennedy committed the nation to "achieving the goal, before this decade is out, of landing a man on the moon and returning him safely to Earth." The Apollo Program cost over $5 billion dollars per year for the following decade. This intensive research and development project, which culminated in Neil Armstrong's "giant leap for mankind" on July 20, 1969, has produced 60,000 "spinoff" technologies for the civilian economy. According to the National Aeronautics and Space Administration (NASA), the estimated value of these technological breakthroughs is $600 million—$700 million per year.

Public's Right To Know

Military programs will realize the most benefit from SDI research. Specific spinoffs may include improved radars and sensors and use of the electromagnetic rail gun against enemy tanks. The civilian space program will be enriched by SDI advances in propulsion and sensing devices.

Grant Loebs, "SDI's Trillion Dollar Dividend," The Heritage Foundation *Backgrounder*, March 21, 1988. Reprinted with permission.

Important nongovernment applications of SDI research will be in medicine, where SDI research advances will be turned into practical products for the treatment and curing of eye and bone diseases, and in the advancement and modernization of American industry through the introduction of high-precision cutting machines and laser tools.

Though the primary purpose of SDI is and must remain the protection of the U.S. and its allies from Soviet ballistic missiles, SDI's technological breakthroughs also can be exploited thoroughly by other branches of the U.S. military and by civilian industry. The Strategic Defense Initiative Organization (known as SDIO), which oversees the entire project, should devise ways to spread the emerging technology as widely as possible while maintaining tight security control over highly sensitive discoveries.

In 1986 the SDIO established the Office of Technology Applications to develop and direct a program to make SDI technology available to other branches of the military and federal agencies as well as private American businesses and researchers. Congress has passed several laws mandating SDI's technology transfer and the SDIO Office of Technology Applications has developed a system to make information on SDI advances available.

Private industry must be able to take advantage of the ever increasing pool of scientific data available through SDI. Congress should relent on its refusal to allow the SDIO to establish a public liaison office. The general public deserves to be educated about the benefits of SDI research and has a right to know that the tax money spent on SDI already is yielding dividends. And the Administration should use the "spinoff argument" more vigorously in the general debate over SDI funding. The strength of this argument is potentially enormous, especially against the unsubstantiated argument that SDI will cost the taxpayer too much.

SDI conducts research into a number of different technologies. Some, such as lasers and particle beams, show great promise as weapons against ballistic missiles but may require ten to fifteen years of research and development before deployment. SDI research indicates that other technologies will be available in the next five years to construct and deploy an effective SDI system at a reasonable cost. The mainstays of such a near-term system would be kinetic-kill weapons (KKW), space and ground-launched rockets that destroy warheads and missiles by crashing into them.

The SDI research office also is making a major effort to develop new types of sensors for detecting, tracking, and monitoring enemy intercontinental ballistic missiles (ICBMs). These tasks of SDIO's Surveillance, Acquisition, Tracking and Kill Assessment progam (SATKA) are among the most crucial and delicate in the SDI system because of the high complexity of sensor systems and the high degree of accuracy needed to guide KKVs to their targets.

"Many of the major breakthroughs achieved in SDI research are transferable to other applications, both military and civilian."

SDI's research efforts have been widespread and ground-breaking. Progess in miniaturization of electronic components, advanced propulsion techniques, surveillance and tracking systems, superconductor capacity, advanced data and image processing, as well as development of new production techniques and materials lead the list of SDI research successes. Many of the major breakthroughs achieved in SDI research are transferable to other applications, both military and civilian.

Because SDI is a military project under the control of the Pentagon, the bulk of the early technical breakthroughs will be carried over into other military programs. The three military services are eager to take advantage of the wealth of new scientific data and practical technological developments generated through SDI research, and SDI-generated technology is already being put into use in weapons design by all of them.

The Air Force

Air Force air defense systems will be able to utilize several major SDI components. The North American Air Defense Command (NORAD) surveillance mission against enemy air attack would benefit directly from SDI's space-based optical and infrared sensors, which could detect and target enemy aircraft accurately as they penetrated U.S. air space. Ground-based anti-aircraft radar systems would benefit from the improved aerial surveillance capabilities provided by such advanced SDI technology as laser radars and new larger phased-array radars. SDI could provide ground-based sensors and kinetic energy weapons capable of tracking, targeting, and destroying encroaching Soviet bombers and cruise missiles.

Specific benefits of SDI-generated technology to Air Force systems would include:

• Improving the capability of NORAD to track enemy bombers approaching North America, using SDI radar technologies to improve the accuracy and coverage of NORAD radar systems.

• Increasing the efficiency of the Air Force and the Air National Guard in intercepting enemy aircraft by applying SDI infrared and optical sensor technology to increase the accuracy of air defense missiles.

• Using SDI guidance and targeting technology to improve the accuracy of U.S. air-to-air and ground-to-air missiles by increasing the precision of infrared sensors on the interceptor missiles.

• Applying SDI surveillance and tracking technology, such as space-based radars, to improve the Air Force's ability to monitor, track, and intercept Soviet missiles, aircraft, cruise missiles, and ground activities at Soviet air and missile bases.

The Army and Navy

The Navy will benefit immensely from SDI technology. The Soviet threat to the U.S. Navy is growing from the increasing numbers of sophisticated Soviet ships at sea and the increased pressure the Soviets are putting on strategic and commercial waterways. To protect itself, the Navy currently utilizes a multi-tiered defense system much like that envisioned for SDI. The tiers consist of defenses appropriate to various distances, which increase in intensity as the threat gets closer to the naval asset being defended. This system depends greatly upon surveillance, communications, and accurately guided interception of hostile missiles and aircraft. Much SDI technology therefore is almost directly transferable to this naval defense mission.

Spinoffs from advances in communications, computer systems, and computer software now under development by SDIO will increase the Navy's fleet defense by tying the various ship-board and airborne components into a more closely knit system able to operate at high speed. SDI sensing systems, meanwhile, can improve undersea surveillance and sonar scanning of enemy submarines. Torpedoes and guided anti-torpedo weapons, as well as anti-mine torpedoes, and naval anti-missile defense systems will be made more accurate by SDI-improved guidance and sensor technology, including laser radars, a carbon dioxide laser capable of scanning objects as a radar does.

Most important to the Army are the potential SDI spinoffs applicable to conventional weapon modernization and anti-armor warfare. SDI's electromagnetic rail gun, for example, shows exceptional promise as an anti-armor gun. Army weapons experts agree that the next generation of Soviet tanks will be very hard to stop with currently available anti-armor weapons. The problem is to find a relatively lightweight weapon capable of penetrating increasingly more formidable Soviet armor on tanks and personnel carriers. The electromagnetic rail gun, under development by SDIO, propels bullets at extremely high speeds by using immensely powerful electromagnetic pulses. It may be the Army's anti-armor answer. The speed of projectiles propelled by rail guns is far greater than can be achieved using traditional rockets and shells, thus increasing their ability to break through Soviet armor.

"New technology generated in SDI laboratories will be made available expeditiously to qualified U.S. companies and individual entrepreneurs."

Other technological benefits to the Army from SDI research include:

• Advances in high-speed computers and computer chips will enhance the capability of tanks to engage more than one target at a time. Typical targets include enemy tanks, trucks, and other vehicles. Advances in computer-controlled systems which enable tanks to fire shots in rapid succession, will enhance the accuracy and speed of armor operations by enabling tanks to shoot at more targets within a shorter period of time.

• The Short-Range Anti-Tank Weapon (SRAW), the Ground Launched Hellfire (GLH) missile system and advanced mine detectors, all initially researched by SDIO, could enhance the Army's battlefield capability against armored forces and mechanized infantry.

• Satellite modernization and radars will improve the Army's satellite communications and the coordination of troop movements on the battlefield by providing accurate information more quickly to commanders in the field.

Civilian Applications

The civilian space program will be the most significant nonmilitary beneficiary of SDI technology. Because so much of SDI's mission focuses on space-based elements such as satellites, space-based

sensors and radars, and space-based anti-ballistic missile systems, SDI scientists and program directors work very closely with the National Aeronautics and Space Administration. Relevant technology is thus spun off very rapidly. SDIO and NASA are working together on a big rocket (heavy-launch-vehicle) capable of carrying into space payloads in the range of 120 thousand to 150 thousand pounds. This big rocket's huge payload is essential to the launching and assembling of the heaviest space-borne elements of SDI, such as laser guns and particle beam accelerators, as well as to the construction of NASA's space station.

In addition to the big rocket, NASA will benefit directly from SDI research into lightweight structural materials such as very strong, metal alloy materials similar to particle board, which can take the place of heavier structural steel, aluminum and graphite. Other dividends include durable lubricants, more efficient rocket propulsion systems, increased miniaturization of electronic circuits and computer circuits, and SDIO-developed superconductors of enhanced capacity.

Through the SDIO Office of Technology Application, and under congressional and presidential mandate, new technology generated in SDI laboratories will be made available expeditiously to qualified U.S. companies and individual entrepreneurs. The SDIO Civil Applications progam will disseminate information on SDI research data through a sophisticated computer data base (the SDI Technology Applications Information System or TAIS). This data base includes information about new and unique SDI-generated technologies and will be available to all qualified government, business, and academic clients who have been approved by the Department of Defense to receive advanced technology information.

Industrial Uses

According to Colonel James Ball, the Director of SDI's Technology Applications Office, new technical breakthroughs made by SDI scientists can be available for industrial use as soon as six months from the time they are made. This very rapid turnaround will contribute immensely to the value of the SDI civilian technology transfer program by allowing new discoveries to be put to use in the civilian economy very quickly.

Several SDI breakthroughs already have been earmarked for industrial use. Examples:

• Diamond Crystal Coating, a new method of efficiently utilizing industrial diamonds by depositing thin layers of diamond on cutting surfaces, will increase the precision of industrial saws. Industrial saws are used to cut a wide variety of patterns for consumer goods and industrial equipment.

• High-temperature carbon fiber ceramic materials for automobile and jet engine components may lead to higher performance, higher efficiency engines.

• The SPOCK supercomputer, a new computer technology that combines hardware, computer code, and semiconductor devices, may revolutionize artificial intelligence by speeding the processing time of computers.

• Lighter, smaller, more capable and energy efficient electrical components being developed by SDI will improve electrical engines, computer circuits, and electrical power supply systems.

• Numerous technologies derived from SDI research will be available to the consumer to improve "The Kitchen of the 21st Century." Possible advances include laser ovens and cooler, more efficient refrigerators and freezers.

Medical Uses

Perhaps the most promising nonmilitary applications of SDI technology are in medicine. The largest and most impressive medical program under SDI is the Medical Free Electron Laser Program (MFEL), which is being conducted at five regional SDI laboratories. This Advanced Free Electron Laser, about 1,000 to 10,000 times stronger than the next most powerful laser, has proved in laboratory tests to be able to vaporize diseased tissue with pinpoint accuracy, leaving the surrounding healthy tissue unharmed. The applications of such a laser are numerous. The precision surgery made possible by MFELs will be used in cancer surgery, eye surgery, bone surgery, and numerous other medical procedures that currently are either impossible or very dangerous for the patient.

"Research . . . is yielding scientific breakthroughs of enormous value to U.S. conventional military programs and to U.S. industry and medicine."

Other significant medical applications of SDI technology include:

• Bioglass, a material developed under SDI to replace human bone without its being rejected by the body, could greatly reduce the danger of bone-graft operations.

• A new, safer, and less complex method of producing radioisotopes used to diagnose brain and heart diseases early and without risk has been developed by SDI researchers.

• A method of cleansing the blood in blood-bank supplies of such viruses as herpes, measles, and HIV—the AIDS virus—has been discovered in SDI labs.

The potential benefits of SDI to private industry are enormous and should be fully exploited. To further enhance the spread of SDI technology throughout the U.S. economy, Congress should allow SDIO to educate the public about the progress of SDI research. In arguing its case for SDI, the Administration should use the spinoff argument.

Conclusion

The Strategic Defense Initiative may offer the U.S. a means of ending the nuclear arms race by eliminating the threat of a successful Soviet first strike. SDI thus deserves support on its merits, as a way to safeguard the U.S. and the Free World from Soviet nuclear blackmail and attack. As SDI scientists pursue this goal, however, their intense and concentrated research also is yielding scientific breakthroughs of enormous value to U.S. conventional military programs and to U.S. industry and medicine. These benefits should be made known to the American public, to help SDI get the enthusiastic public support it deserves.

Grant Loebs is a policy analyst for The Heritage Foundation, a Washington-based think tank.

"The United States cannot afford to devote substantial portions of its technical talent and resources to an SDI megaproject."

viewpoint **16**

SDI Harms the Economy

Council on Economic Priorities

Beyond the dollars-and-cents issues of who pays for SDI [Strategic Defense Initiative] and who will benefit from the effort to develop a strategic defense system lies a broader question: how will an expanding SDI program affect U.S. technological growth and the competitiveness of our nation's high technology industries?

As in earlier debates over space and defense progams, advocates of SDI contend that it will produce a cornucopia of valuable spinoffs that will improve productivity and enhance the competitiveness of the U.S. economy. Critics, on the other hand, argue that the extreme performance goals, relative inattention to cost, and secrecy that characterize this program and other military-sponsored R&D [research & development] efforts will minimize spinoffs to civilian industry. Moreover, many economists and science policy experts maintain that in the highly competitive international environment of the late twentieth century, the United States cannot afford to devote substantial portions of its technical talent and resources to an SDI megaproject and strengthen crucial industries at the same time.

Commercial Spinoffs

Even if SDI proceeds to deployment, consuming hundreds of billions of dollars and employing a substantial proportion of the nation's scientific and engineering personnel, its drain on the economy could be mitigated or even neutralized if the program yields substantial commercial spinoffs that improve the competitiveness and productivity of U.S. civilian industries. The SDI Organization [SDIO] has opened an Education and Civil Applications Office to explore spinoff possibilities and educate the public about these potential benefits of strategic defense

research. In the words of Capt. Chuck Houston of the new office, "the American taxpayer deserves some return that he or she can look at, and see that something other than defense can come out of SDI."

Captain Houston acknowledges that "there aren't a whole lot of applications yet," citing contracts with ten universities to investigate the medical uses of free electron lasers (FELs) as the most promising area so far. Even here, the adaptation for civilian use may take some time. "The technology is still quite new, and has mostly physics applications," says Houston. "The FEL is quite large—not the size of a refrigerator, but the size of a room—and the mobility is very limited. There are only a few of them available in the United States. It's not yet available in stores—you can't just go and pick one up."

Over time, however, former SDIO chief scientist Gerold Yonas argues that the program offers a broad range of potential commercial spinoffs:

> For the civilian economy, there are reasonable expectations that SDI research in lasers, particle beams, sensors, computer systems and software, and hardening will find applications in high strength, high temperature and wear-resistant materials, nondestructive testing, optics and holography, pattern recognition and artificial vision, faster and more sophisticated computers, improved commercial air traffic management, and automated methods for manufacturing high technology components at low cost.

The difficulties of accurately predicting commercial spinoffs of a major R&D effort like SDI are daunting at this stage. Nonetheless, several features of this program and other related military R&D efforts indicate that the likely spinoffs will be relatively limited.

More Productivity?

Government research and development is justified primarily by the assumption that private industry is not likely to invest in activities, such as expanding basic knowledge and protecting the nation, that do not have direct commercial payoffs or are too expensive for one company to undertake alone. By funding basic research and making early use of innovations, government expands the nation's technological base and thus contributes to economic health. Numerous studies, employing a variety of methodologies, have attempted to substantiate this assumption empirically.

The space program has probably been the most carefully analyzed government R&D effort. Motivated in part by a desire to strengthen its position in the yearly budget battle, the National Aeronautics and Space Administration [NASA] commissioned a variety of studies examining the economic benefits of space science. The results of these studies, and the responses they stimulated, clearly bear on our understanding of how SDI will affect the national economy.

In one of the most important of these studies, in 1965 Chase Econometrics examined the relationship between NASA R&D expenditures and productivity growth. Through statistical testing, Chase concluded that NASA R&D was strongly associated with rising productivity between 1960 and 1974. For every $1 million spent by NASA in 1976, Chase predicted productivity improvements resulting from the expenditure would enhance economic output by $23 million in 1984. This is clearly quite a high rate of return on an R&D investment.

The General Accounting Office (GAO) took issue with the Chase study, noting that its conclusion "would indicate an enormous impact on the economy. The literal interpretation is that, if NASA had spent nothing in R&D, productivity would have declined in the United States between 1965 and 1974. That is, NASA R&D is given credit for all of the nation's productivity growth, and more besides." In examining Chase's statistical analysis, the GAO found that the results were very sensitive to the assumptions made. For example, if the time period and variables used in the statistical testing were altered slightly, the relationship between NASA R&D and productivity became weaker and results were statistically insignificant. . . .

One cannot assume a specific government project will either help or hinder private sector technological growth. The burden of proof clearly resides with those who would justify government projects on the basis of their commercial impact. Unless the supporters of a specific project can demonstrate how it will help, we should assume that the effect is at best neutral.

Little Basic Research

How does SDI fit this general pattern of distribution of government R&D activities? A key factor affecting spinoffs from any government research program is the mix of basic and applied research involved in the project. Basic research allows more room for innovation, is conducted in an open environment conducive to the dissemination of research findings, and is generally considered to have more potential for commercial spinoffs. In contrast, applied research is mission oriented, frequently restricts disclosure of its findings, and usually results in fewer commercial spinoffs because of these restrictions.

"One cannot assume a specific government project will either help or hinder private sector technological growth."

SDI research funding has accelerated at a time when the Pentagon already controls a growing share of the nation's R&D funding. Under the Reagan administration, the Pentagon's share of government funding for research has risen from 50 percent in 1980 to 73 percent in 1986. DOD's role is particularly strong in certain key fields critical to high technology growth: more than half of all federal research in mathematics and computer sciences is Pentagon sponsored, as is over 80 percent of research in electrical engineering. At the same time federal support for civilian R&D programs declined by roughly 20 percent in real terms between 1981 and 1985. If SDI continues on its current funding path, federal R&D funding will be tipped further toward military projects.

Current DOD-sponsored R&D is heavily weighted toward development efforts, which are least likely to yield civilian spinoffs. In FY [fiscal year] 1985, 90 percent of DOD R&D was devoted to development of specific weapons systems, while less than 3 percent went for basic research. By contrast, basic research accounts for more than 37 percent of civilian R&D programs sponsored by the federal government, development for only 30 percent. As the Congressional Budget Office has noted, "the predominance of development within the DOD R&D budget has been increasing. . . . The smallest increase has been in the category of 'technology base,' which has the greatest relevance for civilian industrial activities." In effect, the growth of DOD R&D has shifted the entire federal R&D budget away from basic research toward development efforts, to the detriment of the future civilian technology base.

SDI is sure to accelerate this trend. Gerold Yonas has asserted that "about one third" of current SDI funds "will be spent to develop the technology base." If so, SDI would be much more oriented toward basic research than is the average DOD R&D program, although still lagging [behind] the average federal civilian R&D program in this regard.

However, SDI research may already be moving away from this basic research orientation. Independent estimates indicate that 49 percent of the FY 1987 SDI budget request was for experiments. The three program elements that represent the bulk of SDI funding—surveillance, acquisition, tracking, and kill assessment; directed energy weapons; and kinetic energy weapons—have emphasized hardware demonstrations and prototype development in their budget plans for the next five years.

A further factor likely to limit commercial spinoffs from SDI is the program's extreme performance requirements and "gold-plated" technologies. While military research in the post-World War II era helped usher in the age of computers and commercial airliners, the requirements for operations in space are so different from those on earth that there is little likelihood of similar breakthroughs spinning off from SDI research. In fact, civilian spinoffs from military spending have become harder to find as defense moves into space and as weapons become highly complex and expensive. Such overdesign, often the rule rather than the exception, produces little benefit for the cost-conscious civilian market.

Some of the work performed within SDI will have commercial value, for example, very high speed integrated circuits (VHSIC) are being developed for the near real-time data processing required by a strategic defense system. Yet it is hard to imagine private uses of such major SDI technologies as high-energy lasers, particle beams, large optics, or infrared sensors. Hans Peter Durr, director of Munich's Werner Heisenberg Institute for Physics, has raised this question with respect to SDI: "The ability to forge a sword may be useful for making plowshares—but do we need the ability to burn holes in metal at a range of 1,800 miles?"

Too Much Secrecy

The third major obstacle to spinoffs from SDI is the aura of secrecy and security restrictions within which the research is likely to be carried out. Those SDI innovations that do have commercial potential—such as new software languages, next-generation computers, and new integrated circuit designs—may be classified for years to come. The Pentagon has already begun to control the dissemination of scientific findings in the name of national security. This trend will clearly reduce the possibility of spinoffs from SDI. Pentagon restrictions on disclosure of information about VHSIC chips, for example, are so tight that "close-up, head-on photographs of VHSIC chips cannot be published because of concerns that the Soviets might be able to determine the chip architecture and reverse engineer the device from such data."

The Reagan administration has issued new guidelines to federal agencies restricting the release of a broad range of government data that are unclassified but considered sensitive. Already, the DOD has restricted access to several civilian engineering meetings, including the March 1985 meeting of the Society of Photo-Optical Engineers, at which unclassified research on high-energy lasers was being presented. Restraining access to technological advances hinders their commercial development by industry. Manufacturers of VHSIC chips, for instance, may have to obtain security clearance for all employees, visitors, and even some suppliers. Moreover, chip designs used by the military may have to be reworked before they can be used in civilian products. One manufacturer asks, "How much must I change my VHSIC gate-array design to escape security restrictions?" In contrast, programs like those of Japan's Ministry of International Trade and Industry (MITI) and France's European Research and Coordinating Agency (Eureka) program, developed for the purpose of advancing high technology, are unimpeded by security restrictions and benefit from easy diffusion to the commercial sector. While U.S. industry waits for access to this restricted information, France, Japan, and other competitors will be forging ahead with private and civilian programs to sharpen their competitive position in high technology markets.

> *"It is hard to imagine private uses of such major SDI technologies as high-energy lasers, particle beams, large optics, or infrared sensors."*

Finally, there may be "negative spinoffs" from SDI research. That is, some areas of technology or equipment designs may be developed or maintained in civilian use despite flaws in safety, performance, or cost-effectiveness because military research subsidies have pushed development in a particular direction that would not have been pursued if the goal was commercial technological development. John P. Holdren and F. Bailey Green argue that "the pressurized water reactor (PWR), developed to power nuclear submarines and transferred to the civilian sector as the mainstay of the U.S. commercial nuclear power program" is one such negative spinoff.

> The PWR's high power density and very high-pressure coolant—results of criteria appropriate for the submarine application—make it inherently more vulnerable to accidents, earthquakes, and sabotage than are a number of other reactor designs that might have materialized from a strictly civilian reactor program. Tacked-on safety systems motivated by these vulnerabilities have greatly inflated the cost of PWRs and their close relatives, boiling water reactors, and it is possible to suppose that the commercial nuclear

industry would be better off today if it had done without the spinoff of military reactor technology.

While a strict balance sheet cannot be calculated at this early point in the SDI program, the details of its content and structure that are already known suggest it is extremely unlikely to spawn substantial civilian spinoffs, and it will certainly not generate enough commercial activity to "pay for itself." At best, the likely commercial effect of SDI will be comparable to the limited impact of military R&D on the electronics industry, in the view of the Office of Technology Assessment: "Although there have been many examples of secondary and indirect impacts, in no case can *recent* military spending in the United States, France, or the United Kingdom be shown to have stimulated commercial developments in a major way."

Although some SDI supporters claim the program "will lay the foundation for an educational-vocational renaissance for the American labor force," there is serious concern among economic policy experts that the growth of strategic defense research could place the Pentagon in charge of the nation's de facto industrial policy, thereby distorting the entire direction of U.S. high technology development.

"The Star Wars vision will divert attention, leadership, and energy from other pressing national problems."

Robert Reich, industrial policy expert and author of *The Next American Frontier*, estimates that the SDI Organization alone will control roughly 20 percent of U.S. high technology venture capital in the next four years, with potentially disastrous results for U.S. competitiveness. The military will be the principal sponsor of research in key areas such as very high speed integrated circuits, advanced computers, and optics. According to Reich, "The problem is that never before on this scale have we entrusted so much technological development to the Pentagon in so short a time. A handful of Pentagon officials are pre-empting scientific resources and picking winners and losers of the technology race, with large defense contractors advising them." . . .

Wasting Talent

The problem with relying on military R&D to lead the way in technology development is twofold, as Ann Markusen, professor of city and regional planning at the University of California at Berkeley, has argued. First, it condemns important industries like steel, automobiles, and metalworking to technological neglect. Markusen points out that the steel industry "has been denied $15 *million* by the Reagan Administration for its project of 'leapfrog

technology.' U.S. high-tech industries, on the other hand, will enjoy the benefits of $30 *billion* in research outlays under SDI."

Moreover, in those areas that do receive substantial military R&D funds, it is not at all clear that U.S. industry can compete by using spinoffs in areas where Japan and other industrial competitors are investing directly. For example, while U.S. artificial intelligence research is focused heavily on developing battle management systems (for strategic defense and other military purposes) at the expense of "intelligent libraries" and computer-assisted education, Japanese research in artificial intelligence is focused directly at improving business and consumer productivity and improving the delivery of social services.

Perhaps the greatest danger is that the Star Wars vision will divert attention, leadership, and energy from other pressing national problems. Economist Lester Thurow has summed up this argument eloquently: "In many ways, a prosperous civilian economy is our best military defense. The real economic case against Star Wars is that the civilian economy—battered by trade deficits that have no parallel in human history—could well use some of the time, attention, and money now being lavished on Star Wars."

The Council on Economic Priorities is a public service research organization headquartered in New York.

"This mass of legalese is one of the simplest, most radical attempts in history by the leaders of two adversary nations to resolve a point of tension between them."

The INF Treaty: An Overview

Strobe Talbott

The very title of the document is a mouthful—*Treaty Between the United States of America and the Union of Soviet Socialist Republics on the Elimination of Their Intermediate-Range and Shorter-Range Missiles.* It runs to 169 single-spaced typewritten pages, with 17 articles and three annexes. Nearly every word has been haggled over for years. . . .

Yet reduced to its essence, this mass of legalese is one of the simplest, most radical attempts in history by the leaders of two adversary nations to resolve a point of tension between them. Never before has the word elimination appeared in the heading of a nuclear arms-control treaty. It is a dramatic example of the practitioners of nuclear diplomacy taking a sword to the Gordian knot.

There is a short, simple version of how this agreement came about: Once upon a time the man in the White House said to the man in the Kremlin, "Hey, you've got a whole category of weapons we don't like. We've got a whole category of weapons you don't like. Why don't we just wipe clean the slate?" After 72 months of contentious, suspenseful, stop-and-go negotiation, the man in the Kremlin said, "O.K. It's a deal." With that, Mr. Gorbachev comes to Washington, pen in hand.

But before the ink is dry on the last page of the treaty, new disputes are emerging. Some Senators, presidential candidates and West European strategists are saying, Granted, it's the deal we asked for, but is it the one we *should* have asked for? And do we want it now? Two-thirds of the Senators must in effect answer yea for the treaty to become U.S. law. Their answer will depend in large measure on their understanding of the history of how the agreement came about.

And that history is anything but short or simple.

Early in his first term, Ronald Reagan was preparing to give one of the most important speeches of his presidency. He had inherited from Jimmy Carter a perplexing piece of unfinished business: what to do about a new class of missiles that Leonid Brezhnev's Soviet Union had arrayed against Western Europe. Each was mounted on a mobile launcher and armed with three highly accurate warheads that could be fired nearly 3,100 miles. In a minor coup, Western intelligence discovered that the Kremlin's strategic rocket forces secretly referred to this formidable weapon by the innocent-sounding name Pioneer, the Soviet equivalent of Boy Scout or Girl Scout. NATO designated it the SS-20 and warned that it constituted a major escalation in the arms race.

Under pressure from its NATO allies, the Carter Administration had committed the U.S. to the "dual track" decision of 1979. The U.S. would offset the Soviet missiles by deploying a new generation of its own "Euromissiles"—Tomahawk cruise missiles and Pershing II ballistic missiles—while at the same time making a good-faith effort to negotiate with the U.S.S.R. a compromise that would scale back the missiles on both sides.

Left to its own instincts and devices, the Reagan Administration might have abandoned both tracks of the 1979 decision. Assistant Secretary of Defense Richard Perle, the Administration's most forceful and persistent skeptic about traditional arms control, would have preferred to let the intermediate-range nuclear forces (lNF) negotiations languish—the same treatment that was already in store for that other unwelcome legacy with the better-known acronym SALT (for Strategic Arms Limitation Talks). Perle doubted that the negotiating track would lead anywhere and that the West Europeans would have the gumption to follow through on deployment of the U.S. missiles.

But America's European allies were aghast that the

new Administration might renege on the 1979 commitment. They had a friend in court in Alexander Haig, the hard-charging Secretary of State who had been NATO commander in the Ford and Carter Administrations. He made INF a test case to prove that the new President could simultaneously stand up to the Soviets in the military competition and sit down with them at the bargaining table. Haig pushed for a negotiating position similar to that favored by the Carter Administration—fewer Tomahawks and Pershing IIs in exchange for fewer SS-20s.

"The Administration has convinced itself, and now wants to convince everyone else, that the INF treaty is . . . an unprecedented triumph of American persistence."

Haig and other arms-control advocates had two reasons for seeking a deal that would *reduce* missiles in Europe rather than eliminate them entirely: 1) such an outcome seemed realistic and "negotiable," in that the Soviets might accept it; 2) leaving a few missiles in place would reinforce the credibility of the U.S. promise to defend its allies in the event of a Soviet attack.

But the State Department plan was not good enough for the President. It smacked too much of the half-a-loaf compromises of SALT. Reagan told his National Security Adviser of the time, Richard Allen, that he wanted a proposal "that can be expressed in a single sentence and that sounds like real disarmament."

Perle had just what Reagan was looking for: the "zero option." He proposed a straightforward, all-or-nothing package—zero American missiles in exchange for zero SS-20s. That scheme could indeed be presented in a single sentence, which was at the heart of a speech the President delivered on Nov. 18, 1981: "The United States is prepared to cancel its deployment of Pershing II and ground-launched cruise missiles if the Soviets will dismantle their SS-20, SS-4 and SS-5 missiles."

Since then much has changed. Brezhnev and two successors have gone to their graves by the Kremlin wall. All three angrily denounced the zero option as patently one-sided. So did many Western strategists. The U.S. was asking the Soviets to give up real weapons, already deployed at great expense, in return for the U.S.'s tearing up a piece of paper. Washington wags said it was like the Redskins trying to persuade the hated Dallas Cowboys to trade Tony Dorsett for a future draft pick. Administration officials privately conceded that the zero option was not intended to produce an agreement before NATO deployment began in late 1983. Rather, it was a gimmick—part of an exercise in what Assistant Secretary of State Richard Burt, Haig's chief deputy for arms control and Perle's nemesis, called "alliance management"—to make sure the nervous West Europeans kept to the self-imposed deadline. . . .

Even as it prepared to welcome the Soviet leader, the Reagan Administration could not resist the temptation to occasionally gloat over Moscow's apparent capitulation in the face of American steadfastness. Perle has been beaming with the pride of paternity and enjoying the last laugh. The Administration has convinced itself, and now wants to convince everyone else, that the INF treaty is not just an unprecedented accomplishment by the superpowers acting in concert—the elimination of an entire class of modern weaponry—but an unprecedented triumph of American persistence over Soviet intransigence. As Kenneth Adelman, the Perle ally who is outgoing director of the Arms Control and Disarmament Agency, put it recently, "For once we had a negotiation, and the good guys won."

There is some truth to that claim. But it is not the whole truth, and it may not turn out to be the most important truth. The story of the INF treaty is also one of Soviet persistence, Soviet ingenuity and, yes, Soviet success. That is a critical element of any arms-control agreement: both sides must feel they succeeded. The Soviet Union set out to keep American missiles as far from its territory as possible. And [on December 8, 1987] it will sign an agreement doing just that.

The Tula Line

The game . . . began about ten years ago, when Ronald Reagan was a radio commentator and Gorbachev was Communist Party boss for the Stavropol region. That was when the strategic rocket forces started deploying the SS-20s. But that same year, Soviet civilian leaders began to have doubts about whether more and more nuclear weapons like the SS-20 necessarily meant more security and power for the U.S.S.R. The Kremlin initiated a gradual shift in emphasis away from nuclear weaponry to conventional weaponry as instruments of Soviet influence and intimidation, particularly in Europe. In January 1977 Brezhnev gave a speech at a World War II commemorative celebration in Tula, a city south of Moscow. The Soviet leader laid down what became known in the West as the "Tula line." In that speech and subsequent elaborations, Brezhnev said nuclear superiority was "pointless," it was "dangerous madness" for anyone even to seek victory in a nuclear war, and the Soviets needed only nuclear forces that were "sufficient" to hold those of the U.S. in check.

Sufficiency was a word and a concept that had been commonplace among Western strategists for at least a decade. Soviet doctrine seemed finally to be catching up.

It was, as Soviets like to say, "no accident" that in the same month as Brezhnev's Tula speech, Nikolai Ogarkov became chief of the Soviet general staff. Marshal Ogarkov was a controversial choice among the top brass. He had been the top military representative to SALT. The civilian leadership apparently picked him because he too believed in sufficiency, parity and stalemate. He also favored Soviet-American agreements as a means of regulating the arms race.

Ogarkov, however, was no dove. The money saved by relying less on nuclear missiles he wanted to spend on advanced conventional weapons. He did not want those rubles diverted to the beleaguered Soviet consumer economy. He was finally demoted in September 1984. But the new chief of the general staff, Marshal Sergei Akhromeyev, was also a proponent of the idea that enough is enough in nuclear weaponry.

There was, in the Tula line, both good news and bad news for the West. A recognition of the need for nuclear sufficiency rather than superiority was welcome, especially if it meant that the Soviet Union might be coaxed into retiring some of its most threatening weapons. The bad news was that Moscow still seemed bent on increasing its influence in Europe—and on using its huge conventional military strength to do so.

"The first American missiles arrived in Europe in late 1983."

Besides, in Moscow's thinking, the partial denuclearization of Soviet military strategy required the much more thorough denuclearization of the American military presence in Europe. Moscow might be more willing to bargain away some of its own missiles, but it was more determined than ever not to sanction the stationing of new, land-based American nuclear weapons near the Soviet border.

On a number of occasions in the 1950s and '60s, the U.S. and its allies had installed American missiles in and around Europe as equalizers, to make up for the Soviet Union's geographical proximity and the numerical superiority of the Warsaw Pact over NATO. In each case, some combination of American ambivalence, West European anxiety and Soviet neuralgia led to eventual withdrawal of the U.S. missiles. For example, at the height of the Cuban missile crisis in 1962, Khrushchev demanded the removal from Turkey of American Jupiter rockets (ancestors of the Pershing II) in exchange for his agreement to take Soviet SS-4s and SS-5s (ancestors of the SS-20) out of Cuba. Says one of Gorbachev's

advisers: "The resolution of the Caribbean crisis established the principle that we would not threaten you with nuclear weapons from within the western hemisphere. But another principle was established too: We put you on notice that forever after we would regard American land-based missiles on the periphery of the U.S.S.R. as an unacceptable threat to our security."

The INF treaty . . . will leave the U.S. without any ground-based missiles in Europe capable of hitting Soviet territory—and without the right to deploy any such weapons in the future. That is every bit as much a mission accomplished in Soviet policy as the accompanying elimination of the SS-20s is a consummation of Reagan and Perle's original zero option.

The bottom line of the INF treaty in 1987 is Brezhnev's Tula line of 1977.

No Right, No Blessing

It has already become part of the U.S.-fostered mythology of INF that the Kremlin had to be dragged kicking and screaming into eventual acceptance of the zero option, that it was not until earlier this year [1987] that Gorbachev finally seized the long-standing American proposal and made it his own. Here, too, the history is more complex. On Nov. 23, 1981, five days after Reagan first unveiled the zero option, Brezhnev on a trip to Bonn proposed the eventual elimination of all medium-range weapons "directed toward Europe," plus the elimination of all shorter-range missiles.

For Brezhnev then, just as for Gorbachev now, what mattered most was U.S. missiles in Europe that could reach Soviet territory. For two years, from late 1981 until the end of 1983, Soviet negotiators hammered away at the unacceptability of any new American deployments. The head of the Soviet delegation at the talks in Geneva, Yuli Kvitsinsky, then a bright young star of the Soviet diplomatic corps, declared that the U.S. had no "right" to deploy missiles in Europe and the U.S.S.R. would never "bless" the stationing of even a single cruise missile or Pershing II east of the Atlantic.

Kvitsinsky's American counterpart was Paul Nitze, 80, a grand old man of American nuclear strategy. In 1982 they engaged in an extraordinary, one-on-one mini-negotiation—the so-called walk in the woods— that resulted in a tentative deal that would have sacrificed the Pershing II but allowed the U.S. a stripped-down deployment of cruise missiles to counter a residual force of SS-20s. Cruise missiles fly subsonically at low altitudes and are vulnerable to enemy air defenses. The Pershing II ballistic missiles arc to the edge of space and can strike targets inside western Russia in a matter of minutes. The deal was repudiated by both men's home offices. It was shot down in Washington (particularly by Perle) because it meant giving up the Pershing II, and in Moscow because it meant allowing even a few U.S. cruise missiles in Europe.

The first American missiles arrived in Europe in late 1983. The Soviet gerontocracy had painted itself into a corner, leaving no alternative but to walk out in Geneva. There was a widespread assumption in the West, encouraged by Washington, that the battle was over. The U.S. and NATO had won. The Soviets now had to accept the new reality of modern American missiles on European territory.

Not so, says a Soviet official with close ties to the military: "Our generals were more determined than ever to get the American missiles out and to keep them out. The general staff concluded that Brezhnev really blew it by provoking the U.S. into installing the Pershing IIs in the first place and then not having the wit to make a deal to get rid of them."

First, however, there had to be a successor who could do something.

Shortly before Reagan's second Inauguration, in January 1985, Secretary of State George Shultz met Soviet Foreign Minister Andrei Gromyko in Geneva and agreed to get negotiations started again. They settled on a formula for three sets of talks—INF, the Strategic Arms Reduction Talks [START], and a new negotiation on defense and space, focusing on the Strategic Defense Initiative [SDI], or Star Wars. But the Soviets insisted, and Shultz agreed, that the three sets of issues would eventually have to be resolved "in their interrelationship." The Soviets said at the time that this phrase meant hard-and-fast "linkage": there could be no separate deal on INF or START without American concessions on Star Wars. The Americans pressed from the outset for an INF deal that did not require concessions on Star Wars.

At the first session of the talks in March 1985, the chairman of the Soviet delegation, Victor Karpov, a bluff, crusty veteran of SALT, trotted out virtually all Moscow's old demands and added some new ones for good measure. He went out of his way to stress that his plenary statement had been approved "at the highest level"—by Mikhail Gorbachev, who had become General Secretary one day before.

It was the toughest opening bid that experienced Americans could remember. There were dark jokes about canceling hotel rooms and packing for home. However, the head of the U.S. delegation, Max Kampelman, had just the opposite reaction. He could see that he and his colleagues were in for a long haul, but he did not mind. "We'll be talking for a long time," he told Shultz. . . .

The first hint that the game might be changing came in 1985, when the Soviets tipped their hand on two critical points. One was the status of SS-20s in Soviet Asia. The U.S. had been insisting that the zero option must be "global in scope": it must eliminate SS-20s in Asia too, since they are mobile weapons that in a crisis could be moved to threaten Europe.

In May 1985, Gorbachev publicly suggested that his government would be willing to freeze its SS-20 forces east of the Ural Mountains. Shortly afterward the Soviet delegation in Geneva tabled a proposal to that effect. The General Secretary was rapidly becoming his own chief negotiator.

"The other key issue was whether, despite earlier Soviet statements to the contrary, INF might be delinked from an agreement on long-range strategic weapons and Star Wars."

The other key issue was whether, despite earlier Soviet statements to the contrary, INF might be delinked from an agreement on long-range strategic weapons and Star Wars. [Maynard] Glitman took [Alexei] Obukhov [chief American and Soviet negotiators] aside and tried to persuade him of what he called the "logic" of a separate deal on INF. "Let's assume," he said to Obukhov, "that we were to agree fully with the position you've taken on INF. We could see reaching an agreement without linkage. Couldn't you?" Obukhov paused, thought hard, then replied that he could indeed see such a possibility. A few days later, after checking with his superiors, he told Glitman, "I can tell you that my answer was correct." Once again it was Gorbachev who officially enunciated the new Soviet position. On Oct. 3, during a visit to Paris, he said an INF agreement might be possible "outside of direct connection with the problem of space and strategic arms."

Meanwhile Karpov told U.S. negotiators in Geneva that he was "alarmed at how slow things are going." Kampelman, who relished the chance to out-stonewall a master stonewaller, told Kvitsinsky, who was now serving as one of Karpov's deputies, "Yuli, I don't see why Victor is so alarmed." Kvitsinsky replied, "Well, I'm alarmed that you are not alarmed."

Americans sensed that Gorbachev and Eduard Shevardnadze, who had replaced Gromyko as Foreign Minister in July, had decided that INF was the one area where progress might be possible at the first Reagan-Gorbachev summit, which was to be held in Geneva in November. With that event looming, Karpov turned almost plaintive: "We have an opportunity to resolve some important issues in advance of the meeting of our leaders."

Shortly afterward Karpov and Obukhov tabled a new INF proposal that at first blush seemed to capitulate on the most critical issue of all. In what a Soviet official in Moscow later recalled as a "momentous sacrifice that left blood on the floor of more than one ministry," the Kremlin proposed its own version of an "interim agreement": the U.S. could keep a handful of the missiles it had deployed in Europe in exchange for a reduction of Soviet SS-20s in range of Europe and a freeze on those in Asia.

It turned out, however, to be the first in a series of now-you-see-it, now-you-don't Soviet teasers. Moscow's "interim" proposal was the bait for a summit, and it had a number of familiar strings attached. The Soviets had devised a complicated formula that would give them their long-sought compensation for the British and French independent nuclear arsenals that the U.S. insisted should not be part of any INF deal. Also, the U.S. would be allowed to keep only cruise missiles in Europe. The more capable Pershing II ballistic missiles would have to come out. Moreover, the Soviet proposal stipulated that the U.S. would have to commit itself to the eventual elimination of all American missiles in Europe.

At the Geneva summit in November, Reagan refused to yield on the British and French forces and insisted that the U.S. would keep Pershing IIs in West Germany as long as there were SS-20s deployed anywhere in the U.S.S.R. But in their final communiqué, the two leaders agreed there should be early progress toward an INF interim accord.

"[Gorbachev] called for cancellation of Star Wars, a 50% reduction in strategic weaponry and 'complete liquidation' of Soviet and American INF missiles 'in the European zone.'"

After this first summit, Gorbachev was more impatient than ever with the diplomats of both sides who were slogging away in Geneva. He was also emboldened about his ability to compete with the Great Communicator in Washington for the hearts and minds of international public opinion. Said one of his advisers: "The General Secretary decided to take a more active, direct and public role in advancing the process. He resolved to seize the bull by the horns."

He did it in January 1986 with a bold stroke: a proposal for a comprehensive settlement that subsumed all three sets of negotiations. It was a three-stage, 15-year plan for total nuclear disarmament. The first stage called for cancellation of Star Wars, a 50% reduction in strategic weaponry and "complete liquidation" of Soviet and American INF missiles "in the European zone." In Geneva the next day, Karpov opened Round 4 of the nuclear and

space talks with a verbatim reading from the eleven-page Gorbachev proposal. It was marked SEKRETNO even though virtually every word had just been distributed worldwide.

Karpov & Co. once again seemed surprised by their leader's tour de force in public diplomacy. When the American negotiators pressed them for clarification, the Soviets' answers were confused and contradictory—particularly on the critical issue of whether an interim INF deal was contingent on U.S. acceptance of restrictions on Star Wars.

Kvitsinsky told a West German politician that Gorbachev's proposal superseded earlier Soviet willingness, enshrined only two months before in the summit communiqué, to settle for a separate INF treaty. An interim agreement, said Kvitsinsky, was now "impossible." Linkage was again the order of the day.

But not for long. Two weeks later Kvitsinsky was contradicted by Gorbachev himself. The Soviet leader again showed his penchant for going over everyone's head—this time directly to influential American liberals. On Feb. 6, during a conversation with visiting Senator Edward Kennedy, the Soviet leader said an interim INF deal, independent of START and SDI, might indeed be possible. Moreover, such an agreement could be signed at a summit in Washington later in the year.

This latest play of the delinkage card brought broad smiles in Washington. The sweet smell of vindication was in the air.

Some Western analysts, however, had growing doubts about whether delinkage and the zero option would necessarily be an unmitigated blessing. A veteran intelligence official cast a pall over an interagency meeting in February by administering what he called a "heavy dose of reality therapy." Consider, he said, the danger posed by a new Soviet ICBM [intercontinental ballistic missile]—the SS-25, a mobile, three-stage, intercontinental version of the two-stage, intermediate-range SS-20. "Not a single one of the SS-20s that Gorbachev will be giving up can hit the U.S., and not a single SS-25 is affected by an INF treaty. So there's nothing to stop him from replacing every SS-20 he takes out of service with an SS-25 that can hit us easily. What's more, SS-25s can cover the same targets in Europe that the SS-20s have been covering. Given an INF agreement but absent a START agreement, we could end up having more Soviet warheads aimed against us than before and our allies could be in no better shape than they are now."

The chief Sovietologist on the staff of the National Security Council, Jack Matlock (who is now U.S. Ambassador to Moscow), favored the zero option but cautioned against euphoria. Gorbachev's latest tactic, he told colleagues, "might be a breakthrough in the negotiations, but it would also achieve the elimination of American INF missiles in Europe."

As so often happened within the Administration, Gorbachev's offer produced an outbreak of guerrilla warfare. The State Department and the Arms Control and Disarmament Agency lined up behind a counterproposal that accepted elimination of INF missiles in Europe but insisted further on a 50% reduction of SS-20s in Asia. The Pentagon, represented in a series of heated meetings by Perle, wanted to hang tough on "global zero" (zero SS-20s in Asia as well as Europe) and also to force Soviet concessions on their "shorter-range" SS-12/22 and SS-23 missiles.

Nitze, who had become special adviser to Shultz and Reagan on arms control, had never liked the zero option, but he now did his best to sell it to U.S. allies in Europe. During one of his frequent missions, European leaders told Nitze that they had invested considerable political capital in accepting the American missiles. They had withstood domestic opposition by arguing that the missiles were necessary to assure "coupling" between America's nuclear forces and its defense of NATO. It would be awkward to justify the removal of all the U.S. missiles, even as part of a deal that eliminated the threat of the SS-20s. NATO strategy still required an American nuclear "trip wire" to deter a Soviet conventional attack.

"The Americans had now seen Gorbachev delink and relink INF and SDI so often that they calculated it was only a matter of time before he delinked yet again."

As an aide to British Prime Minister Margaret Thatcher put it, "We would have preferred to leave a token deployment of American missiles in Europe. Nitze's own walk-in-the-woods scheme would have been a far better outcome than the zero option from a strategic point of view. If, however, the U.S. allowed itself to be snookered by the Soviets into the damn-fool zero option, then we told Nitze in no uncertain terms that we wanted it to be a version of the zero option that extracted the maximum price from the Kremlin."

Yet the Reagan Administration was reluctant to back away from the zero option, partly because it had been Reagan's proposal to begin with. Glitman instead proposed a modification of the interim solution: an immediate reduction of INF missiles on both sides combined with a schedule for achieving the "global" elimination of INF missiles by the end of 1989. Obukhov replied dryly: "We'll study this more carefully, but on initial consideration, it looks like the zero option."

Meanwhile, there had been a shake-up in the delegation. Kvitsinsky, a specialist on Germany, was transferred to Bonn as Ambassador so he could argue the Soviet case in fluent German against U.S. Envoy Richard Burt. Obukhov moved from INF to START, and his deputy, Lev Masterkov, moved up to be chief INF negotiator. Masterkov had a reputation as an "iron-pants" negotiator of the old school. There was debate among the Americans over whether his appointment meant the Kremlin was indeed ready to move to closure in INF and wanted someone who would get the best possible deal in the final stages, or whether his assignment would be to stall the talks.

The Last 20 Minutes

In September 1986, the Soviets once again began dangling the bait of an INF-only summit. They were, said Karpov, under instructions to take "practical steps" that would assure progress at a "meeting at the highest level." They were prepared to concentrate on the most promising area, which was INF, and, in Karpov's words, to leave START and SDI "off to one side, in hopes of making as much progress as possible on those at the summit itself." They proposed their own version of an interim solution: 100 INF warheads per side in Europe—although with no Pershing IIs—and a freeze on Asian SS-20s.

The Reagan Administration, to the relief of some of its own members as well as numerous Europeans, saw an opportunity to retreat from the controversial zero option and to reinstate the interim solution, with token missile deployments in Europe. U.S. negotiators tabled a response that seemed quite close to the Soviet proposal: each superpower could keep 100 INF warheads in Europe, but with some Pershing IIs permitted.

The next day Foreign Minister Shevardnadze, who was in the U.S. for a visit to the U.N., called on President Reagan at the White House and delivered an invitation from Gorbachev to Reagan for a meeting in Reykjavik. An official on the powerful Central Committee Secretariat, Georgi Kornienko, said in Moscow, "We feel it is important to make progress somewhere, and INF appears to be the only area of opportunity." All indications were that the deal the Soviets had in mind was the interim agreement, not the zero option.

But when Reagan arrived in Reykjavik, hoping to put the finishing touches on an INF treaty, he found himself confronted instead with yet another Gorbachev blockbuster. Gone was the offer of an interim INF agreement that would allow the U.S. to maintain some missiles in Europe for a limited period. In its place was the zero option, which would meet the long-standing Soviet objective of keeping all American missiles off the Continent. As before, having originally proposed the zero option, the Administration felt it could not reject it at Reykjavik.

There was an almost audible sigh of relief from NATO capitals when, at the end of the dizzying weekend, the deal fell apart over the old issue of linkage: Gorbachev made an INF deal conditional on a comprehensive strategic agreement that would confine Star Wars to laboratory research. Reagan refused on the grounds that such limitation would "kill" the program.

"While the U.S. has succeeded in separating INF from the bigger issues of START and SDI, the success could prove temporary and illusory."

The Americans had now seen Gorbachev delink and relink INF and SDI so often that they calculated it was only a matter of time before he delinked yet again. Moreover, it was increasingly clear that he was determined to eliminate American missiles in Europe.

As they prepared for the end game of INF, the Soviets upgraded their Geneva team. Karpov was recalled to Moscow and replaced by a Deputy Foreign Minister and former No. 2 Soviet diplomat in Washington, Yuli Vorontsov. Suave, self-assured and experienced in back-channel diplomacy, Vorontsov proposed spending less time in large sessions, which were, he said, "too polemical." Instead, they should concentrate on the individual negotiations, including working lunches for himself and Kampelman.

But, as before, it remained for Gorbachev to make the next move. In February [1987] over a Friday dinner, Vorontsov dropped a hint to Kampelman that he expected new instructions to arrive soon from Moscow. The next day Kampelman was receiving one of the steady stream of congressional delegations that came through Geneva to look in on the talks. Emerging from a long lunch with the visiting legislators at the U.S. mission, Kampelman found a message from Vorontsov. The Soviet diplomat gave Kampelman a copy of a major statement by Gorbachev that would be released later that evening.

As expected, Gorbachev delinked the INF deal once and for all from the issues of SDI and START. In order to achieve the basic Soviet goal of keeping American missiles out of Europe, he was willing to accept a separate INF agreement along the lines of Washington's original zero option.

For its part, the Reagan Administration became resigned to making the best of the zero option and accepting yes for an answer. Despite the qualms of many about entirely eliminating America's nuclear-missile deterrence in Europe, Reagan remained just as attracted as ever to the "elimination of the entire class of land-based missiles." That bold and simple idea was far more compelling to him than recondite concerns over "coupling" and "extended deterrence," just as it had been when he originally proposed the zero option in 1981.

But there was still much work to be done. "Gromyko used to be fond of saying that the last 20 minutes of a negotiation are the most important," Kampelman told Shultz after Gorbachev's February announcement. "Well, we're entering the last 20 minutes." They lasted nine months.

Kampelman's toughest job was persuading the Soviets to accept a global zero-zero plan: no SS-20s or shorter-range INF missiles anywhere in the U.S.S.R. He explained how such a treaty would help eventually with the politics of ratification in the U.S. Senate. "A big concern of the Senators," said Kampelman, "will be verification. It will be far easier to verify a treaty that achieves a global zero outcome than one that leaves some missiles in Europe or Asia. What we're now talking about would be clean, crisp and far more verifiable than the interim agreement." To underscore the political obstacles that Reagan could face at home, Kampelman showed Vorontsov a newspaper article by Richard Nixon and Henry Kissinger that was highly critical of the prospective treaty.

During a Shultz visit to Moscow in April [1987], Gorbachev made an important concession: shorter-range INF missiles would indeed be eliminated throughout the U.S.S.R. As usual the Soviet team in Geneva was slow to catch up with its home office. Vorontsov at first said that his government was prepared to "zero-out" shorter-range missiles only in Europe. It took some weeks for him to bring his delegation's position into line with what Gorbachev had already told Shultz in Moscow.

The treaty . . . cannot exist in a vacuum for very long. While the U.S. has succeeded in separating INF from the bigger issues of START and SDI, the success could prove temporary and illusory. What the experts, Soviet and American alike, call "conceptual" linkage remains a fact of life. Unless the SS-25 and other ICBMs are dealt with in a strategic agreement sometime soon, they will eventually nullify the good news being celebrated in Washington and around the world.

Strobe Talbott is a diplomatic correspondent. His books include Reagan and Gorbachev *and* Deadly Gambits: The Reagan Administration and the Stalemate in Nuclear Arms Control.

"The conclusion of the Soviet-American INF treaty is the first step in carrying out the programme for the abolition of nuclear weapons."

The INF Treaty Strengthens World Peace

Sergei Golyakov and Yuri Lebedev

Editor's note: Part I of the following viewpoint, excerpted from the Soviet magazine New Times, *is a report by Sergei Golyakov, who attended the Supreme Soviet's deliberations on the INF [intermediate-range nuclear forces] treaty. Part is by Yuri Lebedev.*

I

The scale, and novelty of the decisions the deputies of the Supreme Soviet are to take created a special atmosphere in the marble hall of the Supreme Soviet building, where the first joint meeting of the Foreign Affairs commissions of the two houses of our supreme legislative body was held on Tuesday, February 9 [1988].

Apart from the deputies, the meeting was attended by representatives of the ministries and departments involved in preparing the treaty, foreign diplomats and a large group of Soviet and foreign journalists.

The treaty is the starting point for the restructuring of international relations on the principles of the new political thinking, and the present vision of the world. Work on the treaty took six years. Both sides repeatedly checked and verified the possible consequences for their security, and minutely weighed the political, economic and other pros and cons. The fact that the signatures of Mikhail Gorbachev and Ronald Reagan stand under the treaty testifies that it does no danger to the legitimate interests of either side, and gives no one-sided advantage or superiority. Equality and equal security are the foundation on which it rests.

It was these factors that won the treaty the widespread approval of the world community and the Soviet and American public. At the same time, as Egor Ligachev noted when opening the joint meeting of the commissions, the Soviet people have a number of questions. They want to analyze in

Sergei Golyakov, "Repercussions of the Summit: The Only Option," *New Times*, no. 7, February 1988.
Yuri Lebedev, "A Realistic Programme of Action," *New Times*, no. 3, January 1988.

detail what the treaty gives them and what the world gains from it, how it will affect the security of the Soviet Union.

What Are the Guarantees?

In working on the process of ratification, the legislators have to take into account all shades of public opinion, answer all questions and allay all doubts. The members regard this work of explaining and convincing the public as an integral part of the procedure.

What worries people most? What guarantees are there that the United States will abide honestly and consistently by the treaty?

Senate hearings in the United States showed that, alongside a desire seriously to determine the essence of the treaty, there are a number of attempts to disparage it, present it as "deceit," as a "red trap," and torpedo its ratification. There are people in the West, some of them in very senior posts, who seek to prove that only nuclear weapons can save the world, and that they are the panacea for all wars and conflicts.

And on these grounds they seek to use the signing of the treaty on the elimination of medium- and shorter-range missiles as an excuse for the further buildup and modernization of the nuclear arsenals of the United States and NATO as a whole in compensation.

Former U.S. Ambassador to France Evan Galbraith tried to persuade senators that the best that could be done was to preserve the missiles (medium- and shorter-range) and then try to persuade the world that it is in the interests of Western security, as it will deter the Soviet Union. As for the Supreme Commander of NATO forces in Europe, General John Galvin, although he believes the Senate should ratify the treaty, he insists on the modernization of NATO's forces both non-nuclear and tactical nuclear. The Senate hearings have shown that the principal enemies of the transition from confrontation to cooperation were and remain suspicion, prejudice, mistrust, and outmoded dogmas of prenuclear thinking.

Soviet Concern About US Compliance

What the Soviet people are particularly concerned about are the attempts of the American administration to circumvent the ABM [antiballistic missile] treaty and produce offensive space weapons. But such an approach would rule out a 50 per cent reduction in strategic offensive weapons. Such a cut is possible only given strict adherence to the ABM treaty.

The Soviet side proceeds from the premise that after its ratification the Soviet-American treaty must not merely be strictly observed, but should make a start in the process of disarmanent right across the board. It should and could pave the way for a 50 per cent cut in strategic offensive weapons, banning nuclear tests, reducing tactical nuclear and conventional armaments, armed forces, and banning chemical weapons. "We believe," Ligachev said, "that the conclusion and implementation of the treaty cannot fail to be accompanied by the expansion and deepening of international economic, scientific and cultural relations, and in general, as we see it, by an improvement in Soviet-American relations. All people everywhere would benefit from this. But one can't expect the course of disarmament to run smoothly of its own. There are powerful forces in the United States and Western Europe that stand firmly for the militarization of the economy and public life. So the efforts of parliamentarians in all countries and peace forces the world over will be required."

"The treaty cannot fail to be accompanied by . . . an improvement in Soviet-American relations. All people everywhere would benefit from this."

One group of questions is prompted by the image of the other partner to the agreement, by a desire to visualize his intentions and understand his mood. Another arises when people think of the security of their own home. Those who still remember two world wars find it hard to agree to even a partial dismantling of the shield that protects our borders. Such people cannot be suspected of favouring nuclear weapons, but neither can their experience be ignored.

People also have misgivings about the difference in the number of missiles to be eliminated (the other side's number is only a fraction of ours), about the

British and French nuclear potentials that remain untouched, about America starting production of new types of chemical weapons, no less destructive than nuclear arms. Some are worried by the decision to eliminate Soviet medium-range missiles in Asia. There are no American missiles there.

No Grounds for Anxiety

Foreign Minister Eduard Shevardnadze and Defence Minister Dmitry Yazov told the meeting firmly that there are no grounds for anxiety. All the equations in the treaty have been carefully calculated. They do not upset parity. The balance of security remains, only at a lower level.

The ministers' views were listened to very carefully and noted. But the members who spoke after them, while paying the Soviet-American treaty its due, urged their colleagues to take another and closer look at the document, and carefully weigh all its main points.

Democracy implies the right to have your own opinion, to have detailed and full information on any issue, including foreign policy matters. After all, the questions of war and peace bypass no home, no family. So they have to be resolved together, so that no one should feel he has been discounted. It is the duty of members to ensure that. That is why they suggested considering the document from the point of view of the questions raised by the public. To this end it was proposed to set up a preparatory committee made up of members of both houses. The proposal was adopted. The preparatory committee will get down to work straight away. It will hear the views of specialists and experts, and then present its own findings to the deputies.

That meeting was only a beginning. The bulk of the work remains to be done. One might at that point put a full stop, were it not for a letter from one of our readers in Kolomna, Dr. Alexei Kurganov. He writes about the atmosphere in which the new Soviet-American treaty became possible, the shifts in our way of thinking that made the Washington agreement a reality.

To Build Peace Together

Alexei Kurganov writes: "I often remember my grandfather's stories about how we and the Americans met on the Elbe, having defeated the Nazis. My grandfather wasn't on the Elbe himself. But I remember well how he used the words 'we' and 'the Americans.' 'We' meant all our people, all of us. 'The Americans' also implied the entire nation. So it was not only the soldiers of the two countries who met on the Elbe, but two peoples." That was the beginning of our friendship.

"But after that we started slipping. The cold war, [the Great] Powers, anti-communism elevated to the rank of state policy. They called us 'Russian bears,' and we retaliated with 'world gendarme.' That lasted for a long time. My hand itches to write that the blame falls entirely on them. But should I? One can hurl accusations at each other endlessly, but is it worth it? Thank God we have begun to realize that making mutual accusations is not the best way of conducting a dialogue. Then it became clear that such a situation could not continue. Stability was essential.

"The three main steps to such stability have been made: the meetings between Mikhail Gorbachev and President Reagan in Geneva, Reykjavik and Washington. They are all links in the same chain, with the gold link, of course, being the meeting in the American capital. The treaty on the elimination of medium- and shorter-range missiles is not just a military or political document. It is a treaty affecting all of mankind. Looking at it, I feel that my grandfather's attitude to Americans is beginning to gain the upper hand. All right, we are different, but we have one thing in common—we live together on the same planet. And we should never forget that for a moment."

"We have only one choice—to build peace together."

That too is a voice from the people. Clearly it is the voice of a young man, not weighed down by the bias and prejudices of the cold war, a young man capable of sober assessments. That brought him to his main conclusion: we have only one choice—to build peace together. That should be remembered by the legislators when they pronounce their verdict on the document that should become the first step towards a nuclear-free future.

II

The staged elimination of all nuclear weapons by the end of the century, known in the West as the Gorbachev Plan, is characterized by political realism and does not diminish the security of any nation.

The Reykjavik summit showed for the first time that the complete abolition of nuclear weapons is, in principle, a feasible proposition. It became evident that the Soviet Union and the United States can give up nuclear rivalry and confrontation. Though such an agreement was not reached in Reykjavik, the summit marked a turning point on the road to a nuclear-free world.

Subsequent developments have brought changes in the Soviet programme. In a joint effort, Moscow and Washington found a link in the chain of nuclear weapons which they thought would be easier to break. On December 8 [1987] the two countries signed an agreement on the elimination of medium- and shorter-range missiles.

The conclusion of the Soviet-American INF treaty

is the first step in carrying out the programme for the abolition of nuclear weapons, a historic landmark on the road to further disarmament. For after the ratification of this treaty, nuclear missiles of two classes, several thousand units of deadly weapons, will be destroyed. Three continents will get rid of a part of the monstrous nuclear burden. True, it is a small portion of the nuclear arsenal. What is important, however, is that part of the nuclear weapons will be eliminated and, it must be hoped, a new stage will open in the process of real nuclear disarmament.

Also important is the fact that the Soviet-American INF treaty promotes further results in the mutual efforts towards cuts in strategic offensive weapons. The Soviet programme for ridding the world of nuclear weapons concentrates on a reduction and, finally, the complete elimination of strategic offensive armaments. This was a key issue during the talks in Washington between the leaders of the two powers.

Strategic Offensive Arms

So far the two countries have discussed a 50 per cent reduction in strategic offensive weapons. The result would be to retain no more than 1,600 delivery vehicles (missiles and bombers) and 6,000 warheads. It is obvious that hundreds of missiles and thousands of nuclear warheads would be destroyed by both sides. What is also important is that a fresh, strong impulse would be given to the nuclear arms reduction process. The nuclear arms race, in its most threatening field of strategic armaments, would be hampered and the world would become a safer place to live.

An agreement on drastic cuts in strategic offensive weapons would make it possible to include other countries in the elimination of nuclear armaments, release funds badly needed for mankind's development, and facilitate progress towards ending the race in other classes of weapons.

Of course, one should not underestimate both the objective and subjective difficulties of the forthcoming talks. The scale of a reduction is itself a psychological barrier which will be invisibly present in the course of drafting an agreement.

But the two countries already have practical experience as regards agreements in such a sensitive field as nuclear armaments. Definite progress has also been made in mutual relations. This gives hope for success. The leaders of the two states have declared that they will spare no effort to bring about such an agreement. . . .

It is obvious that a reduction of strategic offensive weapons calls for measures to stop attempts to circumvent the future agreement through other channels of the arms race, mainly through outer space. That is why the Soviet and American leaders have instructed their delegations to work out an agreement that would compel both sides to observe the ABM treaty within an agreed period of time.

Many difficult problems also have to be solved in such a delicate sphere as verification. There may arise problems in the development of Soviet-American relations, the maintenance of strategic stability, and lessening the risk of nuclear war at a lower level of nuclear confrontation. The solution of these and other immediate problems should become a new stage in the implementation of the Soviet programme for the complete elimination of nuclear weapons in the world.

A Concrete Plan of Action

The main thing is, however, that the programme for the elimination of nuclear weapons—initially a political document of intent—has in the two years since January 15, 1986, become a concrete plan of action. Not simply a workable plan, but a plan which is already being put into practice.

Sergei Golyakov is a political analyst for New Times, *a weekly Soviet magazine. Yuri Lebedev is a major general in the Soviet military.*

"The most likely effect of new agreements will be to reduce the Free World's ability to respond to Soviet threats and defend against them."

viewpoint **19**

The INF Treaty Endangers World Peace

International Security Council

The Free World, and especially its leading power, the United States, is the author of its own disarray. For more than twenty years, behind the screen of the arms control process, the Soviets have engaged in a massive buildup of both strategic offensive and defensive arsenals to the point that, today, they possess enough ICBMs [intercontinental ballistic missiles] with the necessary combination of accuracy, reliability, and warhead yields to eliminate the vast majority of land-, sea-, and air-based U.S. strategic weapons in a surprise attack—with enough held in reserve to constitute a powerful incentive for the United States not to retaliate and thus put the U.S. civilian population at risk. Even more, by placing a large part of the U.S. strategic deterrent in jeopardy, the Soviets might feel free to employ conventional military force at lower levels, in confidence that the United States will back away from confrontation. Finally, the Soviets have the world's only true strategic defense program, consisting of ballistic missile defenses, thousands of surface-to-air missiles, radar capacity well beyond that formerly planned for the entire U.S. Safeguard ABM [antiballistic missile] program, and R&D [research and development] programs for new generation ABM systems more long-standing, more lavishly funded, and in many respects more advanced than U.S. SDI [Strategic Defense Initiative] programs.

Negative Effects of INF

Yet today the United States is about to rush into a new arms control accord with the Soviets, largely unverifiable, that will have the net effects of a) further diminishing the credibility of the U.S. strategic deterrent; b) increasing the decisive impact of Soviet conventional superiority, already

International Security Council, "International Security Council Strategic Assessment: 1988," *Global Affairs*, Winter 1988. Reprinted with permission.

overwhelming; and c) sowing discord within the core Western alliance, NATO, thereby inviting the spread of both European neutralism and U.S. unilateralism. Focus on this new INF [intermediate-range nuclear forces] agreement tends, moreover, to confirm the logic of anti-nuclear movements worldwide: It erodes public understanding of the foundations of nuclear deterrence (and public support for essential new programs) and suggests that nuclear weapons themselves—even defensive nuclear weapons—are the principal threat to peace and security, rather than the expansionist goals of the Soviet Union....

The United States and its allies cannot allow the promise of arms control to delay or dilute efforts to stabilize the nuclear balance. For some twenty years, the Soviets have used this lure to restrain allied arms modernization programs and to facilitate their own drive for nuclear superiority. The nuclear imbalance today testifies to their success.

Soviet Strategy

The reality is that, since the beginning of the strategic arms limitation talks, the Soviets have developed and deployed the very force capabilities that were to have been prevented by arms control. Now they see prolongation of the process as a means to maintain and expand their nuclear superiority, to prevent the United States and its allies from developing and deploying the defenses that would offset Soviet advantages, and to preserve their monopoly in ballistic missile defenses.

Current Soviet initiatives represent a continuation of this strategy: using the hope of arms control to generate unilateral constraints on Free World arms programs, the SDI in particular. As in the past, these initiatives promise *future* reductions in Soviet capabilities in return for *present* constraints on U.S. and allied programs. But also, as in the past, once the constraints are achieved—with the virtual certainty that in democratic societies they will not easily be removed—the Soviets will use them to expand their own advantages.

The Soviets may be willing actually to reduce, or even eliminate their advantage in a particular nuclear weapons category in return for political and strategic gains that more than compensate. In the INF context these include: the weakening of NATO's political cohesion; the unraveling of NATO's strategy of flexible response; increased pressures in Europe for "denuclearization" (which in effect will be unilateral: all other Soviet nuclear forces remain in place); a new period of détente, which would dull allied vigilance and competitiveness while providing the Soviets with the benefits of Western trade, technology, and financial credits; and self-imposed political constraints on U.S. and allied arms modernization.

"While the Soviets have been building up their nuclear forces, the United States and its allies have reduced their own."

Moreover, the Soviets may expect that they can achieve such gains by promising arms reductions without actually carrying them out. All experience counsels caution in accepting the current Soviet proposals at face value.

• The Soviets' public emphasis on disarmament has created in the Free World the illusion that Soviet military programs are slackening. Yet, in fact, Soviet nuclear arms programs, both offensive and defensive, continue to grow. While they speak of eliminating their SS-20 force, the Soviets are expanding forces capable of substituting for the SS-20—forces of both shorter and longer range. And, while the Soviets have been building up their nuclear forces, the United States and its allies have reduced their own. NATO, without regard to ongoing arms control negotiations or the buildup of Soviet theater nuclear forces, has reduced its nuclear forces by 2,400 warheads in recent years—an actual reduction for which NATO's leaders have received scant credit even from their own publics.

• Arms limitation agreements with the Soviets never have accomplished what was expected by Western negotiators and publics at the time of their conclusion. SALT I was supposed to prohibit Soviet substitution of heavy ICBMs for light ICBMs. It did not do so. All such agreements, at Soviet insistence, have contained loopholes and loose definitions that have created the illusion of limitations but have left the Soviets technically free to circumvent them. Ambiguities are routinely included in arms control agreements precisely to mask an actual lack of

agreement. An INF accord, predictably, would be no exception.

• Even in spite of such permissiveness, the Soviets still selectively violate the agreements. President Reagan has formally found, and so reported to the U.S. Congress, the Soviet Union to be in violation of every major arms control agreement it has entered into. There is a pattern of Soviet violations, he concluded, that "undercuts the integrity and viability of arms control."

Looser Verification Provisions

Yet, astonishingly, the most recent administration reaction to Soviet violations was to *loosen* previously established verification requirements. The serious problems with verifying the agreements currently being negotiated make it very likely that significant Soviet violations could occur over a period of time without being verified. Moreover, given verified Soviet violations of existing agreements, it is imperative to note that verifiability and *enforcement* are very different matters. The administration should at a minimum insist on strict Soviet compliance with existing agreements before signing new ones, and disassociate itself from those agreements the Soviets already are violating.

The most likely effect of new agreements will be to reduce the Free World's ability to respond to Soviet threats and defend against them, and to create opportunities for the Soviets to expand their strategic superiority with a monopoly on strategic defenses and little concern over allied countermeasures. Arms control agreements reflect the reality of the strategic balance: Because that balance today is unsatisfactory, arms control agreements based on it will be unsatisfactory. Firm U.S. and allied measures to improve the strategic balance, including ABM deployment, must *precede* further agreements with the Soviets.

The INF Agreement

This agreement, as it seems to be taking shape, exemplifies the fundamental problems of arms control. It is being presented as a victory for the United States and its allies because it does call for the removal of a larger number of more sophisti-cated Soviet forces in exchange for fewer U.S. forces. But this claim is fraudulent:

1) The agreement will be largely unverifiable. The United States simply does not know how many launchers and warheads the Soviets have, deployed and in storage, and it is highly unlikely that any verification scheme can be devised to overcome this shortcoming.

2) Even if it were, the Soviets—as already noted—cheat, and the United States has been singularly lax in forcing a correction of past cheating.

3) The concept underlying the agreement is strategically unsound. NATO today finds itself hard

pressed defending against Soviet chemical, biological, conventional, and tactical nuclear superiority. Accordingly, it has adopted a strategy that relies ultimately on a nuclear response to a Soviet attack that cannot be repulsed conventionally. The deterrent effect of this nuclear threat has been at the heart of NATO strategy—and this is precisely the reason the Soviets are so eager to enter into an agreement to remove the Pershing IIs facing them, which have the range to hit strategic targets deep in Soviet territory, quickly and with great accuracy.

4) The agreement . . . already has achieved one principal goal: It has created strains between the United States and its NATO allies, a long-sought and much valued goal of Soviet policy.

The Soviet Advantage

Even should the Soviets proceed with the elminination of their SS-20s, the coverage of that force is more than handled by other Soviet capabilities: by substantial excess ICBM capability, including refire capabilities, and by a very large and still growing number and variety of short-range nuclear systems with adequate target coverage for Western Europe— all of which puts particular military, political, and psychological pressures on the Federal Republic of Germany. Elimination of the SS-20 force would not relieve NATO of this type of nuclear threat; in contrast, removal of the U.S. INF would eliminate the only real capability in Western Europe that can threaten significant military targets deep in Eastern Europe and the western Soviet Union. It also would eliminate a force essential to NATO's strategy of flexible response—one that has acquired enormous political and psychological importance in Europe. Elimination of the INF also would impose greater demands on U.S. strategic forces generally, and on NATO's conventional capabilities, at a time when these balances greatly favor the Soviet Union. . . .

"Removal of the U.S. INF would eliminate the only real capability in Western Europe that can threaten significant military targets deep in Eastern Europe and the western Soviet Union."

To safeguard against the risks of the proposed INF agreement, including the likelihood of Soviet cheating, at least four courses of action are required: 1) the United States must move rapidly to improve the strategic nuclear balance, through offensive modernization and deployment of ABM; 2) in Europe, defenses against the principal Soviet attack threats must be established by improved air defenses and deployment of ballistic missile defenses; and 3) by deployment of enhanced radiation weapons, which the alliance earlier agreed would be necessary to counter Soviet conventional attack capabilities; and finally 4) NATO must strengthen its conventional forces.

Members of the International Security Council include former senior military officers, diplomats, government officials, political scientists, economists, and historians.

"President Reagan's 'historic' agreement exposes . . . all of Western Europe to the lengthening shadow of Soviet conventional might."

The INF Treaty Weakens Western Europe's Defense

Josef Joffe and Michel Tatu

Editor's note: Part I of the following viewpoint is by Josef Joffe. Part II is by Michel Tatu.

I

Since 1981 the United States and the Soviet Union have been trying to work out disarmament à deux—in outer space, in the strategic arena down below, and in the European theater, where U.S. cruise missiles and Pershing IIs are arrayed against the notorious SS-20s and a whole family of lesser SS designations. Unable to agree on the two big-ticket items—strategic offense and defense—the superpowers are about to strike a "historic" deal (the "double-zero") on the Euromissiles. . . . Two entire categories of nuclear weapons will disappear: those in the 1,000- to 5,000-kilometer range and those in the 500- to 1,000-kilometer range.

The Soviets recently dragged a half-forgotten hybrid system to center stage—the 700-kilometer "Pershing Ia," owned by the West Germans but with warheads in American custody. The Soviets began to portray those 72 outworn missiles as the last "insurmountable" obstacle in the talks. It was a nice way to test the strength of Chancellor Helmut Kohl's center-right coalition and the resilience of Bonn's American connection—and the Soviets won. Facing "gentle persuasion" from Washington, indifference from Paris and London, and a loudly anti-nuclear Socialist opposition, Kohl caved in. The joint U.S.-German Pershing venture, in place for a quarter century, will go once the two superpowers finish dismantling their Euroarsenals.

Though an INF [intermediate-range nuclear forces] agreement will eliminate only about three percent of the world's stockpile, the appellation "historic" is

perhaps no exaggeration. It would be the first time two great powers actually agreed to *scrap* weapons, and not just to limit their growth (as in SALT [strategic arms limitations talks]) or to demilitarize a particular piece of real estate. It would also be the first time that the swords-into-plowshares business would be effected by voluntary and bilateral contract. Historically there has been plenty of enforced disarmament, imposed by the victors on the losers, and of unilateral arms cuts whereby nations decide on force reductions for their own reasons.

Nevertheless, even if the entire Soviet Euroforce goes to the scrap heap, the United States will gain no security. None of these 600-odd missiles is capable of hitting the American homeland. The Russians, on the other hand, would get rid of 424 U.S. Pershing II and cruise missiles (the planned total was 572), all of which can reach into the Soviet Union. True, the Soviets lose some nuclear options, but help is already on the way. They are about to deploy the mobile, variable-range SS-24 with ten warheads apiece. Officially classified as "intercontinental," these missiles fall outside the scope of an INF treaty. Yet with a range between 3,000 and 9,000 kilometers, they can be launched as easily against Bonn as against Boston.

The Western Europeans are not amused by what they see unfolding in the aftermath of Reykjavik. What may be a sideshow for the two great powers happens to be the main event for those, especially the nuclear have-nots, who must share the continent with a very big neighbor to the east who will always command a surfeit of nuclear weapons. Nor are the Europeans assured by the "Big Twoism" that brought us double-zero. Right after Reykjavik, French prime minister Jacques Chirac invoked the oldest of Western Europe's traumas: once more, "decisions vital to the security of Europe could be taken without Europe really having any say in the matter."

Josef Joffe, "The INF Fallout: Cruisin' for a Bruisin'" *The New Republic,* October 5, 1987. Reprinted by permission of THE NEW REPUBLIC, © 1987, The New Republic, Inc.
Michel Tatu, "Europe's Limited Sovereignty in a Superpower Standoff," *Manchester Guardian Weekly/Le Monde English Section,* September 27, 1987. Reprinted with permission.

The Europeans brought this upon themselves. In 1981 West German chancellor Helmut Schmidt pressured President Reagan to offer the original "zero option"—removal of all intermediate nuclear forces from Europe—in order to get badly needed relief from his anti-nuclear tormentors in the streets and in his own Socialist party. Given the wave of anti-nuclear (and in part anti-American) pacifism then sweeping Western Europe, the "zero option" looked like a wonderful ploy: Would any Russian leader be so stupid as to let go of his great Euro-strategic advantage?

Six years later, with American INFs in place in Europe, [Mikhail] Gorbachev accepted the 1981 offer. Hoping to stop the momentum, the Western Europeans, and the West Germans above all, retorted: "What about your shorter-range missiles, where you have an intolerable monopoly?" Again, Gorbachev's "stupidity" triumphed. He told them: "You can have those, too." It was an offer nobody now dares refuse.

"Nuclear weapons in Europe, especially those that could pierce the Soviet sanctuary, represented the core of Western Europe's defenses."

On the face of it, Gorbachev's offer looks great: what is known in the trade as "asymmetric reduction." He is willing to yield 1,300 warheads for a mere 424 deployed by NATO. Why not take the missiles and run? Answer: because if Gorbachev wants to give away so much for so little, then his idea of a loss and a gain must obviously be different from ours. Needless to say, Gorbachev knows what he's doing.

The Catches

First, in offering double-zero, he invited NATO to get rid of its most modern weapons, the Pershing II and cruise missiles, and to forgo deployment on the next level down. The alliance will thus have to fall back on its nuclear-equipped aircraft, most of which cannot make it beyond Poland and all of which run the risk of being destroyed by a pre-emptive strike before takeoff. If they do get off the ground, they may still not make it to their targets; Warsaw Pact territory is the most lethal air-defense environment in the world. By contrast, there is no effective defense yet developed against Pershing II and cruise missiles.

The second catch is conceptual. Nuclear weapons in Europe, especially those that could pierce the Soviet sanctuary, represented the core of Western Europe's defenses. They were installed to counter a natural Soviet advantage and a congenital Western

European weakness: Russia's preponderance as a nearby superpower and the half-continent's inability (or unwillingness) to field the men and matériel for a conventional defense.

In other words, Gorbachev has proposed not to start with the *basic* imbalance, the geographical-conventional one, but with precisely those weapons deployed to *neutralize* that imbalance. And thus Gorbachev's new thinking turns out to be not so different from the old. "Denuclearization" has been the watchword of Soviet policy since the beginning of NATO. The Soviets fought the insertion of tactical nuclear weapons in the 1950s, and unleashed a massive campaign against NATO's "two-track decision" of 1979 that led to the deployment of Pershing II and cruise missiles. Moscow's nuclear options do not depend on this or that Euromissile. Meanwhile, the drawdown of Western nuclear weapons will unshackle Russia's advantage in troops and tanks, aircraft and artillery.

The third catch is psychological. In the age of "parity" consecrated by the first SALT treaty of 1972, the Europeans have always sought safety in nuclear arrangements that obliterate the distinction between local and global war. Pershing II and cruise missiles standing in the path of a Soviet advance might just go off (whereas a Minuteman III stationed in Montana might not), destroying along with Kiev any dream of a war neatly confined between some Central European "firewalls."

Germany's Fate

Take out long-range INF, and the idea of a war that begins and ends in Central Europe is no longer as absurd as it once was. The shorter the ranges, the deader the Germans who are the prize and the pillar of Europe's postwar order. Which is why an unwritten law of NATO states that Germany must not be what geography has condemned it to be: the venue and victim of East-West war in Europe. Hence the United States has endlessly sought to reassure the Germans with dispositions that threaten to blur the distinction between regional and global war. Yet by leaving in place 4,600 nuclear weapons mainly destined to explode in Germany, double-zero deepens the most powerful "contradiction" within the alliance. In their classic nightmare the Germans play host and target for weapons that will devastate Germany only, and that nightmare presents diplomatic opportunities for the Russians that hardly need belaboring.

To restore equilibrium, the Germans will have to reduce the reasons the Soviets might have to threaten them. The name for this used to be "appeasement." Interestingly enough, Chancellor Kohl's rightish coalition partner, Franz Josef Strauss's Christian Social Union, has already run this theme up the flagpole. Once there are "zones of differential security," a position paper warns darkly, the

"Alliance will lose its meaning. Inevitably, this will engender a reorientation of German foreign policy." Thus the right joins the left on a common platform of neutralist nationalism.

President Reagan's "historic" agreement exposes the West Germans, the holders of the European balance, to a separate nuclear threat and all of Western Europe to the lengthening shadow of Soviet conventional might. From Gorbachev's point of view, not a bad political prize for the price of a few hundred expendable missiles.

II

Let's go back to the Middle Ages briefly and imagine two overarmed fortresses held by two rival princes. Hoping to stabilise, if not to scale down, their quarrels, the two sovereigns agree to keep their long-range weapons which are capable of smashing each other's fortresses, along with everything lying this side of them, of course. But they abandon their smaller artillery pieces, those which allow them to move into the intervening no man's land and threaten each other's fortresses.

Historians would have found such an agreement interesting because it is original, but not of any great significance. For the two princes have not really made peace with each other. They have preserved the possibility of sending their armies into the intervening territories and, with the power of their crossbows, keeping the various principalities scattered in the neighbourhood under their heel, or under their protection.

It is a very minor change for these principalities, with one small difference though: as one of the fortresses is on a distant island and the other is surrounded by a land mass, it means a bigger change for the former than for the latter. The "continentals" in fact continue to occupy their ground and threaten the rest, while the "islanders" abandon all thought of using their continental possessions to threaten their rivals. It is one way of signalling that the fate of these possessions should not be an obstacle to a stable and codified relationship.

"It is these 'bad' weapons that are going to be kept and the 'good' ones that are going to be dumped."

"I threaten you with my big guns, but show you consideration by removing my smaller artillery," says one. "I understand," replies the other, "I'm also withdrawing my smaller artillery." But this, in any case, does not threaten the islanders.

This is roughly what George Shultz and Eduard Shevardnadze have prepared for us . . . pending the signatures of Ronald Reagan and Mikhail Gorbachev. Let it be said at once, for the benefit of those who rather naively imagine all arms are bad and any kind of disarmament good, that the treaty on intermediate nuclear forces (INF) is a positive one. But for the others—and they are many of them all the same, especially in Europe—the agreement is militarily unsatisfactory, even if it seems politically inevitable. On the other hand, some good could still come out of it, if it provided Europe with a chance for asserting its personality and solidarity.

Deterrence

The accord is militarily less than exhilarating for one primary and simple reason. At a moment when Soviet-American strategic parity is the dominant fact of international relations, it should have seemed important to strengthen the deterrent cover that the United States is supposed to give Europe by increasing such deterrence and bringing it closer.

American Pershing 2s and cruise missiles were in fact deployed to make their utilisation more credible because they could be precisely determined in view of their location—not at sea like so many others, but on land, on the very sites likely to be attacked, which placed not only the Soviet command, but also NATO, before a dilemma that military men dread in connection with any weapons system: the dilemma of making use of it or losing it.

The American INF offered another major advantage: they were aimed exclusively at the Soviet Union, that is the territory of the one country with the means for launching an attack on Europe. Consequently, they were infinitely more valuable than the thousands of so-called tactical weapons (very short-range missiles) that the United States has deployed in Europe over the past 30 years whose targets could only be Germans of East and West, Poles, Czechs, French and a few others who have nothing to do with the decision, or are even the "desirable" victims of an aggression. Yet it is these "bad" weapons that are going to be kept and the "good" ones that are going to be dumped.

Similarly today, American determination to resort to nuclear weapons if Europe is attacked is still theoretically guaranteed by the United States' arsenal of intercontinental missiles. But Washington is now rid of the dilemma of making use of its arms or losing them, rid also of the inexorability stemming from this for its commitment. The US president could always press the nuclear button, but he will have an even freer hand for not doing it. And the risks that the Soviet military staff must take into their calculations diminish accordingly.

In Moscow's Interest

The fact that Gorbachev has agreed to make a maximum of concessions to reach this point takes nothing away from this assessment of the whole operation. Far from it. The obstinacy with which Brezhnev, Gromyko and Soviet military chiefs defended their SS-20s for years was probably due much more to unwieldy bureaucracy than to strategic design. On the contrary, it was in Moscow's enlightened self-interest to abandon weapons that were militarily superfluous (the USSR's 14,000-odd nuclear payloads, 10,000 to 12,000 of them strategic, are more than ample for "looking after" Europe in addition to the United States).

Saying . . . that by this means we simply will be returning to the pre-1977 situation when there were neither SS-20s nor Pershings, does not quite answer the question.

"The agreement . . . has in fact the regrettable effect that it could foreshadow the creation of a status of 'limited sovereignty' for Europe."

Right up until the mid-'70s, US superiority was decisive where so-called central strategic weapon systems were concerned. The credibility of the American commitment in Europe hinged above all on this superiority, which is why there was hardly any talk of decoupling. It was precisely because Soviet-American parity, together with the strengthening of the Red Army in all the other spheres, made it necessary to heighten the visibility, proximity, and credibility of deterrence that Helmut Schmidt sounded the alarm in that celebrated speech of his in London in 1977. The SS-20s were still not uppermost on his mind; and it was later, because they offered an infinitely handier alibi to the Allies' position that these missiles became the peg on which NATO hung its 1979 decision. . . .

Political Considerations

Now, Reagan wants an INF treaty not only to draw attention away from his setbacks in the Iran-Contra scandal and go down in history as a "peacemaker," but also to satisfy a powerful public antinuclear movement that has been operating in the United States for a good ten years or so.

As a matter of fact, it was the two superpowers themselves under Nixon and Brezhnev which inched simultaneously towards a less and less nuclear conception of conflicts. . . .

Never have ideas of European military solidarity, of a "pooled defence of Europe," been so fashionable. And this is a direct result of the meeting of superpower minds on the issue of intermediate-range missiles in Europe. . . .

The agreement . . . has in fact the regrettable effect that it could foreshadow the creation of a status of "limited sovereignty" for Europe. Unlike the two superpowers, this continent would be entitled to certain kinds of arms, not to others, and its initiatives in the military sphere, and possibly in other spheres later, would have to fall within the framework of a superpower consensus, not outside it.

This is why everybody feels in a vague sort of way that the agreement under preparation should be paralleled by a European initiative adapted directly to the event. A gesture costing little in substance but significantly as a matter of principle could be for several European countries, at any rate the bigger ones, to issue a formal statement taking note of, or even explicitly approving, the Soviet-American agreement, while reserving the rights and interests of West European states, which are not signatories of the treaty but are immediately concerned. Such a statement would say in effect that the INF treaty covers only Soviet and American arms. It could not be held against European states, whether they are nuclear powers or not. In particular, it could not be utilised to block the possible future constitution of a European defence, even in the area covered by the treaty.

An Impetus for European Defense

In other words, if one day an updated version of the aborted '50s project for a common European defence were to take shape, it should be able to provide itself with the military resources of its own choosing, including those intermediate-range missiles the Soviets and Americans will have abandoned by then. The whole difference between us and them lies in the fact that the superpowers can easily do without such arms which are in no way necessary for their defence, while we cannot see what else other than these missiles in fact could underpin an ultimately united Europe's posture of genuine deterrence.

Author Josef Joffe is foreign editor of Sueddeutsche Zeitung, *a daily newspaper in Munich. Michel Tatu writes for* Le Monde, *a daily newspaper in Paris.*

"Elimination of these Soviet weapons would reduce an important threat to NATO."

The INF Treaty Does Not Weaken Western Europe's Defense

Graham Allison and Albert Carnesdale

The debate on INF [intermediate-range nuclear forces] began in 1977 when West German Chancellor Helmut Schmidt called on the United States to counteract the rapid Soviet buildup of SS-20 medium-range nuclear missiles in Europe. After initial hesitation by the Carter administration, NATO agreed in 1979 to pursue the dual-track strategy: beginning in 1983, a total of 572 U.S. Pershing II and ground-launched cruise missiles (GLCMs) would be deployed in Western Europe; meanwhile, the United States would negotiate with the Soviets to limit the NATO deployments in return for a limitation on SS-20s. At the time, the prospects for successful deployment by NATO looked dim. Most European governments were lukewarm at best in their support of deployment; public majorities in all the countries opposed any new deployment of nuclear weapons in Europe. In the final days of the Carter administration, few members of the national security community believed that the odds of NATO's deploying any medium-range missiles were better than one in three.

As the Peace Movement in Europe mounted increasing pressure against any such deployment, President Reagan countered in 1981 by proposing what came to be known as the "zero option": NATO would cancel deployments in exchange for the elimination of all SS-20s. The primary audience for this American proposal was the European public. The Soviets rejected the proposal out of hand. Western arms control experts dismissed it as incredible and argued instead for the acceptance of an agreement with unequal limits, allowing the Soviets to keep an INF advantage. The Geneva talks stalled. A concept was discussed in 1982 (in the spirit of the earlier Nixon-Brezhnev "Walk in the

Woods") that would have reduced warheads on Soviet SS-20s to 225 while the United States cancelled the Pershing II and limited cruise missile deployment to 300 warheads. But that idea was ultimately rejected by both Washington and Moscow, and in November 1983, in the midst of extensive peace demonstrations, NATO INF deployments began. The Soviet negotiators walked out of the Geneva talks.

Double-Zero

While NATO governments struggled to deploy despite persistent and heavy domestic opposition, the Soviets refused to bargain on INF unless NATO stopped deployment. Still, in 1985 the Soviets returned to the conference table. In January 1986 [Mikhail] Gorbachev picked up the zero option as part of a comprehensive arms control proposal that linked INF with strategic arms reductions and limits on America's Strategic Defense Initiative (SDI). At the October 1986 summit in Reykjavik, agreement on the zero option appeared possible, and collapsed only when Gorbachev held the INF deal hostage for effective limits on SDI research. In subsequent negotiations the Soviets agreed to decouple INF from SDI, to extend the zero option to shorter-range nuclear missiles, making it a "double-zero" deal, and to make the deal global rather than restrict its coverage to Europe. . . . As currently understood, the double-zero arrangement would:

- "zero out" U.S. and Soviet intermediate-range nuclear missiles globally, eliminating 922 Soviet warheads and 316 U.S. warheads in Europe, and in addition eliminating 513 Soviet warheads targeted against Asia (for which the United States has no counterpart).
- "zero out" shorter-range nuclear missiles globally, eliminating several hundred warheads deployed with Soviet SS-12 and SS-23 missiles and an ongoing deployment program, and no

Graham Allison and Albert Carnesdale, "Can the West Accept *Da* for an Answer?" Reprinted by permission of *Daedalus,* Journal of the American Academy of Arts and Sciences, *Futures,* vol. 116, no. 3, Summer 1987, Cambridge, MA.

U.S. launchers (since the United States has no such missiles, no current program for acquiring any, and no agreement with any European country to permit their deployment).

Several collateral features of the agreement deserve special note:

- British and French independent nuclear forces would remain unconstrained. These forces now include about 200 strategic warheads, and are being modernized and expanded to include about 1,200 by the year 1995.
- The five U.S. Poseidon submarines and 150 F-111 bombers currently assigned to Western Europe would remain in place. Together, they can deliver 850 nuclear warheads; both have ranges similar to the to-be-dismantled Pershing IIs and GLCMs. In addition, the United States has more than 500 nuclear-capable aircraft in Europe, and the Europeans have about 1,300. The battlefield nuclear weapons (having ranges of less than 300 miles) of both NATO and the Warsaw Pact would, of course, be unaffected by the agreement.
- Soviet SS-23 missiles would be included among the shorter-range weapons to be eliminated, even though the Soviets had previously argued that they did not belong in this category.

Imagine that the terms of this proposal were reversed: NATO was trading 922 warheads for 316 equivalent Soviet weapons; NATO was eliminating an entire category of shorter-range nuclear systems that included over 120 already deployed in return for nothing on the Soviet side; and Soviet allies were permitted to keep independent strategic forces, scheduled to expand from about 200 to about 1,200 warheads, for which NATO had no counterpart. The likelihood of such terms proving acceptable to an American president, the U.S. Senate, the leaders of the allied European nations, or the public in any of the NATO countries would be remote. Why, then, the outcry—at least in certain quarters? . . .

NATO's Reaction

America's NATO allies, in their initial reaction to the prospect of an agreement to eliminate medium-range missiles, vividly illustrate the power of the final rule of thumb ("If the agreement requires significant change, I'm against it"). Considering the initial European public opposition (some violent) to the proposal to place medium-range missiles in Europe, it is extraordinary that those weapons are now accepted as part of the landscape—part of the new status quo. Whatever the initiative, the dominant initial public reaction of our European allies is one of opposition—whether the proposal is to deploy new nuclear weapons or eliminate weapons already deployed, to increase the number of American troops stationed in Europe or decrease their number, to add new chemical weapons to Europe or

eliminate old ones. The reason for their opposition is not only a desire to counter America's irrepressible penchant for problem-solving, even when dealing with insoluble problems. Rather, our European allies value the status quo precisely because it has nourished four decades without war, following four centuries in which war was more often the rule than the exception.

"Our European allies value the status quo precisely because it has nourished four decades without war, following four centuries in which war was more often the rule than the exception."

Other considerations intensify the attachment of many Europeans to the newly deployed Pershing II and cruise missiles. First, European governments and strategists have invested eight years of political capital in the deployment of NATO's INF. To overcome strong domestic opposition, they developed as many arguments as they could for the support of these weapons. They maintained that the new missiles were necessary to guarantee the credibility of the American nuclear commitment, essential to assure that the Soviet Union would be deterred from any attack on Europe, required to reassure Europeans about the credibility of NATO's defense posture. Strategic analysts understood that many of the technical military arguments were thin. For example, in the early stages of the discussion, the fact that INF in Europe would allow an American president to order a missile attack against the Soviet homeland without directly involving the American central strategic systems led certain strategists to claim that this would have the effect of decoupling U.S. and European security rather than coupling it. But at this stage, having made their vehement arguments in favor of the installations, the European leaders cannot reverse course and argue the contrary without suffering cognitive dissonance and political discredit.

Psychologically Reassuring

Second, the weapons have become symbols of reassurance. Europe's psychological security rests not on its ability to defend itself, but on the expectation that deterrence will work. Deterrence relies on the Soviet leaders' belief that the costs of an attack will vastly exceed any benefits they might hope to attain. Central to this belief is the Soviet expectation that NATO will fulfill its commitment to use nuclear weapons to meet a successful Soviet conventional attack. In the 1970s, as the Soviet nuclear arsenal reached parity with that of the United States, the credibility of America's threat to uphold that commitment despite the risk of Soviet nuclear retaliation on the American homeland began to erode. Since 1982, "no first use" advocates, including former secretary of defense Robert McNamara, have raised further doubts about the credibility of America's nuclear commitment by arguing that "it is difficult to imagine any U.S. president, under any circumstances," fulfilling that commitment in defense of Europe. Europeans have convinced themselves that NATO's INF is the linchpin of all credible deterrence; who can say that Americans and Russians are not convinced of the same? Elimination of nuclear weapons could upset American perceptions, Soviet perceptions of American perceptions, and Europe's peace of mind. . . .

Military Implications

The disadvantages of the INF deal have been asserted most strongly by those concerned with its military implications. Some contend that either "the proposal would be a significant step toward denuclearizing Europe" or that it would make Western Europe "safe for conventional war"; it is also claimed that the agreement would "guarantee that West Germany was the battlefield in a nuclear exchange," "leave a large gap in the capability for flexible response," and "[leave] our allies more vulnerable to Soviet conventional forces and chemical weapons."

Each of these arguments is overstated or misleading, and misrepresents the likely direct effects of the zero-zero agreement. While eliminating NATO's medium- and shorter-range missiles would certainly leave the alliance without some of the weapons it would otherwise have, the INF arrangement would also eliminate a much larger number of weapons on the Soviet side. When one examines the proposed changes in our forces in relation to the proposed changes in Soviet forces, it appears that the net impact would be minor and uncertain, but more likely to be positive than negative.

With respect to the NATO forces, the agreement would turn back the clock to 1983, just before NATO INF deployment began. In the previous decade NATO had no medium-range missiles, nor did it feel any need for them prior to the dual-track decision of 1979. With the INF accord in effect, NATO would face about 900 fewer Soviet nuclear warheads on medium-range missiles and several hundred fewer warheads (including reloads) on 120 launchers for shorter-range missiles. In the new posture, NATO would continue to rely on a strategy of "flexible response"; that is, combined reliance on conventional weapons and nuclear forces for deterrence and defense.

It is difficult to identify any military function that could no longer be performed after the elimination of NATO's medium-range nuclear weapons; no target now vulnerable to destruction by NATO's INF missiles would be invulnerable to destruction by other European-based nuclear weapon systems. Unlike INF missiles, aircraft carrying nuclear weapons face substantial Soviet defenses. But air- and sea-launched cruise missiles are able to deliver nuclear payloads, and are not constrained by the agreement. NATO could, if it chose, increase its forces of air- and sea-launched missiles to cover any crucial shortfall. And while the Pershing II missiles have a capacity to destroy hard (i.e., relatively resistant to nuclear explosions) targets promptly, the deployment of the Trident II (D-5) submarine-based missiles assigned to NATO will duplicate that function.

Still a Credible Deterrent

The credibility of NATO's nuclear deterrent would be unaffected by the agreement. There is no reason to believe that a presidential or NATO decision to use nuclear arms would depend on whether the weapons would be delivered via land-based ballistic missile, sea-launched cruise missile, or nuclear-capable aircraft. If the location from which the warhead is launched or the ownership of the delivery vehicle were judged important, warheads would be available to meet any such requirement (e.g., from German soil or non-German Europe or sea; via aircraft or missiles belonging to the United States, Germany, or some other European nation).

"While eliminating NATO's . . . missiles would certainly leave the alliance without some of the weapons it would otherwise have, the INF arrangement would also eliminate a much larger number of weapons on the Soviet side."

The effects of these changes on Soviet incentives to strike NATO preemptively appear to be small but positive from a NATO perspective. The Soviet Union has been particularly concerned about the Pershing II missiles based in Germany because of their accuracy and short flight time. Removing those missiles would reduce the likelihood of Soviet preemption in time of crisis. . . .

Finally, the Soviet Union has a significant advantage over NATO in its conventional weapons that should neither be forgotten nor minimized. This advantage would not be affected by the INF agreement. To the extent that NATO relies on nuclear weapons to deter a conventional Soviet attack, the reduction of NATO's nuclear arsenal by less than 10 percent will be of little significance. Moreover, the nuclear weapons being eliminated are not the ones that would be preferred for use against Warsaw Pact conventional forces.

While giving up any weapon in our arsenal, especially one invested with symbolism, would constrain some of our military actions, require some changes in our military plans, and stimulate claims that compensating replacements are necessary to fill defensive gaps, it appears that the objective military impact of eliminating NATO's INF would be small.

"Many ties bind the United States and Europe; the deployment of forces like Pershing IIs or GLCMs are thin threads in that complex web."

But no purpose is served by evaluating an agreement only in terms of its effects on the U.S. side of the military balance. The primary benefits for the United States derive from the agreement's effects on Soviet forces. NATO has been rightly concerned about the threat of more than 1,300 Soviet warheads on mobile, accurate SS-20 missiles. These weapons could be effective in a preemptive nuclear attack on nuclear warhead storage bunkers and airfields. Thus, elimination of these Soviet weapons would reduce an important threat to NATO. Similarly, elimination of the Soviets' shorter-range missiles, which could be armed with either nonnuclear (conventional or chemical) or nuclear warheads that would be very effective against NATO airfields, ports, storage sites, and command centers, would also reduce a significant threat to NATO. On balance, elimination of this Soviet conventional capability would marginally lessen NATO's conventional disadvantage. Finally, removal of the Pershing II missiles would undermine the public Soviet rationale for deployment of antitactical ballistic missile (ATBM) systems in Soviet Europe—systems that the United States believes might be upgraded to defend against intercontinental ballistic missiles (ICBMs) and submarine-launched ballistic missiles (SLBMs). . . .

Alliance Cohesion

Claims have been made that this agreement would denuclearize Europe, decouple the United States and Europe, and fuel distrust between the United States and Europe—creating, according to Nixon and Kissinger, "the most profound crisis of the NATO alliance in its forty-year history." Such claims are overstated.

The small proposed reduction in NATO's nuclear forces would leave the West with over 5,000 nuclear weapons in Europe, excluding America's strategic arsenal. Whether this step would be followed by further reductions in nuclear weapons, or even by the denuclearization of Europe at some distant future date, is a reasonable question. But this step in itself would have minimal bearing on that eventual outcome.

The argument about decoupling has become almost theological. As noted previously, at the early stages of the deployment of NATO's INF, many in Europe feared that deployment would decouple American security from that of the Europeans. With U.S. nuclear weapons in Europe, it was argued, some American president might imagine that a nuclear war could be fought in Europe alone (including Soviet Europe) without ever touching the American homeland. U.S. proponents of deployment argued that, on the contrary, the deployment of INF would serve to couple the U.S. and Europe. It would seem that at least one of these arguments is wrong.

The salient fact is that many ties bind the United States and Europe; the deployment of forces like Pershing IIs or GLCMs are thin threads in that complex web. This does not mean that changes in such forces can be made casually. Careful assessment of the military balance before and after an agreement, examination of the agreement's potential effects in particular scenarios, and review of the proposal's benefits and costs in all the relevant currencies are essential. Moreover, meaningful consultation with our European allies constitutes the only approach that will allow a change in forces to strengthen rather than undermine alliance cohesion.

Graham Allison and Albert Carnesdale are deans and professors at the John F. Kennedy School of Government at Harvard University. Allison is also director of the Council on Foreign Relations and a special adviser to the US secretary of defense. Carnesdale is a consultant to the US Departments of State and Defense and the Arms Control and Disarmament Agency.

"The INF Treaty strengthens U.S. and allied security."

viewpoint **22**

The INF Treaty
Improves US Security

George Shultz

The President has forwarded to the Senate, for its advice, the treaty on the elimination of U.S. and Soviet intermediate-range and shorter range missiles. The treaty defines intermediate-range missiles as land-based systems having ranges between 1,000 and 5,500 kilometers. It defines shorter range missiles as land-based systems with a range between 500 and 1,000 kilometers. It requires the United States and the Soviet Union to eliminate their intermediate-range missiles within 3 years and their shorter range missiles within 18 months. Neither side can possess such missiles after they have been eliminated, nor can they produce or flight-test them.

The treaty contains a memorandum of understanding on data giving the locations, numbers, and characteristics of each side's intermediate- and shorter range missiles. It also includes a protocol that sets forth the detailed procedures for eliminating missiles, launchers, support structures, and support equipment. Finally, it contains a protocol that gives the detailed procedures for a variety of inspections associated with the treaty. . . .

The treaty is the result of our steady, patient approach to U.S.-Soviet relations, on the basis of realism, strength, and dialogue. We got this treaty because we were persistent at the bargaining table and because our allies went forward with deployments. The way we and our allies successfully met the Soviet INF [intermediate-range nuclear forces] challenge shows that tough-mindedness, clarity of purpose, and resolve pay off. The INF experience offers important lessons on how to proceed as we confront other challenges to our security.

The INF Treaty strengthens U.S. and allied security. It enhances international stability. It may be

opening a new chapter in arms control—the beginning of reductions. It reduces nuclear weapons, rather than setting guidelines for their future growth. It achieves U.S.-Soviet equality by eliminating substantially more Soviet weapons than American ones. It accomplishes the goals we and our allies set for ourselves 8 years ago, when the deployment of the SS-20 led to the NATO dual-track decision. In short, the treaty is an achievement—as our allies proclaimed—"without precedent in the history of arms control.". . .

Historians may come to see the INF experience as one of NATO's finest hours. We may never know what decision-making process the Soviet leadership went through before deciding to deploy the SS-20. The Soviets may have misunderstood how we and our allies would act during a period of so-called detente. They may have misled themselves into believing that the West no longer would do what was necessary to maintain a strong deterrent, whatever the provocation. If so, they were wrong, just as they were wrong when they calculated the political and military costs before they marched into Afghanistan.

From the late 1950s, the Soviet Union had deployed intermediate-range missiles. These are systems which could threaten most or all of NATO Europe but not reach the United States. The United States had deployed no such missiles since the mid-1960s, although some NATO military officers felt such deployments to be necessary. . . . In arms control terms, at any rate, these systems were not considered "strategic." Thus, they were excluded from the arms control limits in the SALT [strategic arms limitations talks] process.

Soviet SS-20s

In 1977, the Soviet Union began to deploy the SS-20. It was a substantial improvement over its predecessors. It had longer range, greater accuracy,

George Shultz, "The INF Treaty: Strengthening U.S. Security," a statement before the Senate Foreign Relations Committee, U.S. Department of State, *Current Policy*, No. 1038.

nuclear arms/103

and enhanced mobility. Moreover, it had three independently targetable warheads, where previous systems had only one. Both its range and its deployment pattern soon made it a threat, not just to Europe but to all the Soviet Union's neighbors. By our count, 140 SS-20s, with 420 warheads, had been deployed by the end of 1979. When President Reagan and General Secretary Gorbachev signed the INF Treaty in December 1987, the Soviets—according to their own figures—had deployed 405 SS-20s, with 1,215 warheads.

NATO political leaders and military authorities carefully assessed this new threat and consulted extensively on how to counter it. They were concerned that, if left unmatched, this new Soviet capability could lead Moscow into believing that it could intimidate the alliance or even cause the Soviet Union to miscalculate the risks of aggression.

"This agreement has the most stringent and comprehensive scheme of verification in the history of arms control."

To prevent this, in 1979, NATO foreign and defense ministers adopted what has come to be called the "dual-track" decision. On one track, the United States would begin in 1983 to deploy 572 single-warhead intermediate-range missiles in the United Kingdom, Italy, Belgium, the Federal Republic of Germany (F.R.G.), and the Netherlands. There would be 464 ground-launched cruise missiles, which would be deployed four to a launcher. There would be 108 Pershing II ballistic missiles, each on a separate launcher. At the same time, on a second track, the United States would attempt to negotiate limits on U.S. and Soviet INF missiles at the lowest possible level. . . .

Looking back over the negotiations, I think we can divide them into three periods.

From the opening of talks in 1981 until the Soviet walkout in 1983, it is clear that Moscow was simply not willing to bargain seriously or in good faith. The Soviet goal was to block U.S. deployments while retaining their monopoly of INF missiles. They proposed variations on a freeze, which would have left the SS-20s in place and kept us at zero.

The second phase began when the Soviets returned to the talks in 1985. They accepted the reality of U.S. deployments but were still insisting on terms for a treaty that we and our allies could not accept. Sometime in mid- or late 1986, they realized this wasn't going to work.

The third phase began with the important progress the President made on INF at the Reykjavik summit.

The Soviets finally agreed on many of the essential basic conditions for an equitable treaty. Wrestling the issues to the ground from then on, however, was a major effort. But we succeeded, including on the all-important verification issues. . . .

The President's Criteria

We established the following criteria for an INF agreement, which the President set forth.

• There must be equality of rights and limits.

• The negotiations must deal with U.S. and Soviet systems only.

• Limits must be global, with no transfer of the threat from Europe to Asia, or vice versa.

• There must be no adverse effect on NATO's conventional defense and deterrent capability.

• An agreement must be effectively verifiable.

From the outset our position also called for constraints on Soviet shorter range missiles, in order to enhance the effectiveness of the treaty. . . .

Negotiations

The meeting between the President and General Secretary Gorbachev in Reykjavik in October 1986 was the real breakthrough for the INF talks. Moreover, we made important advances in START [strategic arms reduction talks] and began to develop an approach to deal with the need for predictability as we move forward on strategic defenses. Reykjavik set the agenda for the future. From then on, it has been clear that the watchword for the future of arms control is reductions, structured to enhance stability. . . .

Following Reykjavik, the first hurdle to clear was one the Soviets kept moving on and off the negotiating track: linkage with other areas of arms control. Mr. Gorbachev had again made INF hostage to Soviet efforts to cripple SDI [Strategic Defense Initiative]. We and our allies simply refused to accept this, and in February [1987], the Soviets fell off this demand. In doing so, however, they began to equivocate about keeping shorter range missiles in an INF agreement.

In March 1987, the United States tabled a new draft treaty based on the progress made at Reykjavik and containing detailed verification provisions. After a brief recess, the Geneva negotiations resumed in April and continued without recess until the treaty was concluded in December. The pace of work accelerated. In April, the Soviets tabled a new draft treaty. On June 1, the U.S. and Soviet sides agreed on an initial joint draft treaty text. It was heavily bracketed, reflecting continuing areas of disagreement, but it formed the basis of the final treaty. . . .

The hard work of our delegation in Geneva was supplemented by several meetings between me and Minister Shevardnadze. Together with our senior experts, we addressed remaining INF issues as well as the rest of the President's four-part agenda. Depending on whose capital they occurred in, either the President or Mr. Gorbachev were directly involved in these meetings as well, so they, in effect, provided a kind of way for the heads of state to talk to each other.

In INF, there were basically three key tasks remaining:

• Dealing with the issue of the Federal Republic of Germany [FRG], Pershing IA missiles, an issue which the Soviets had elevated to sudden prominence;

• Reaching agreement on the details of effective verification; and

• Ensuring that all of the agreements reached were faithfully and clearly reflected in treaty text.

Cooperation and Verification

We met our objectives in all three areas.

On the first point, the treaty completely protects our position of principle that there be no limits on third-country forces and no impact on existing programs of cooperation with our allies. Our program of cooperation with the Federal Republic of Germany on their Pershing IAs continues. Chancellor Kohl announced last August [1987] that if an INF treaty entered into force, and if the agreed timetable for eliminating INF missiles were adhered to, F.R.G. Pershing IA missiles would be dismantled with the final elimination of U.S. and Soviet INF missiles. At that point, in accordance with a sovereign German decision, our program of cooperation will have ended. Only then will the U.S. reentry vehicles now associated with the German Pershing IAs be subject to the elimination procedures of this treaty.

On verification, you will be hearing in great detail from other witnesses. But let me stress that this agreement has the most stringent and comprehensive scheme of verification in the history of arms control. We worked out our verification regime thoroughly in Washington, with full interagency participation.

We also consulted closely with our basing country allies, whose territory will be directly involved. It took time, but it was well worth it. When the Soviets were finally willing to discuss verification, we were ready. They did so on the basis of thoroughly considered U.S. proposals.

The treaty's verification provisions will ensure our ability to monitor treaty compliance with confidence. They include onsite inspection. "National technical means" are essential, but we simply are not willing to go back to the days of SALT II, when they were all that was allowed. One of the most important elements of the INF Treaty is the precedent it establishes in this area, including onsite inspection.

The structure of the verification regime has built-in redundancies. This sort of double-checking is what gives it its power. One layer of the regime builds on another, to provide a whole that is greater than the sum of its parts. Among other things, the treaty provides for:

• A detailed exchange of data, updated periodically, on the location of missile support facilities, the number of missiles and launchers at those facilities, and technical parameters of those systems;

• Continuing notification of movement of missiles and launchers between declared facilities; that happens once the treaty comes into force;

• An initial baseline inspection to verify the number of missiles and launchers at all facilities declared in the data exchange;

• An elimination inspection to verify the destruction of missiles and launchers;

• Close-out inspection to verify that treaty-prohibited activities have ceased at each of the declared facilities;

• Short-notice onsite inspections for 13 years at declared and formerly declared facilities;

• The right to monitor the Soviet SS-25 final assembly facility at Votkinsk around the clock, to ensure it is not being used for SS-20 assembly; and

• Enhancement of national technical means—specifically, six times a year, the Soviets must, on 6 hours notice, open the roofs of those SS-20 garages that are not subject to onsite inspection, in order to show that no SS-20s are concealed within, and must display the SS-25 launchers in the open.

In addition to its security advantages, the "zero option" makes verification easier and more certain. Once the elimination period is over, the existence of any intermediate-range or shorter range missile would be a violation.

"We and our allies made clear to the Soviets that an INF treaty had to meet NATO's security criteria. We held out for a good treaty, and we got it."

We can never know for certain that a banned missile is not concealed somewhere in the Soviet Union. But the treaty does not permit the Soviets to maintain essential infrastructure for banned missiles or to conduct flight tests. Both testing and infrastructure can be detected by national technical means. Without them, the systems simply become obsolete.

Finally, there is the question of the fine print. It is all there for you to see. There are no secret understandings. . . .

Achievements of the Treaty

It took years of tough bargaining to get where we are today. But the allied position prevailed. It did so because we and our allies made clear to the Soviets that an INF treaty had to meet NATO's security criteria. We held out for a good treaty, and we got it.

The treaty reaffirms the principle of equality. There can be no other basis for U.S.-Soviet arms control.

Because the Soviets had deployed more, the treaty also establishes the principle of asymmetrical reductions. The Soviets will eliminate deployed missiles and launchers capable of carrying about four times as many warheads as those eliminated by the United States.

"Both sides will be reducing deployed intermediate-range systems from the outset, but we will keep a substantial force of Pershing IIs until well into the final months of reductions."

It keeps third-country systems and programs of cooperation with our allies completely out—essential precedents for future negotiations.

It does not rely on Soviet good will but rather requires the most comprehensive verification regime ever, including onsite inspections.

It eliminates the SS-20 threat which directly led to the dual-track decision.

It does not limit U.S. aircraft—which make a critical contribution to NATO's defense.

It strengthens deterrence by significantly complicating Soviet attack planning. From now on, they are denied options that for nearly 30 years they have been able to rely on in any consideration of an attack on NATO.

Moreover, by banning shorter range missiles, it also improves NATO's conventional military posture, by eliminating systems which could target nuclear, conventional, or chemical warheads against NATO's ports and airfields.

Compliance

I want to take a few minutes to address head-on one question which I am sure is on your minds. . . . The Soviet record of compliance with treaties is far from perfect. What does this mean for the INF Treaty?

The basis for negotiations in the first place is that the parties involved seek to agree on an outcome that is in their mutual interest. The Soviets would not have assumed the obligations in the INF Treaty if they had not found its terms satisfactory.

But we have also learned from experience. We have incorporated into the treaty some lessons we learned the hard way. Its terms are more detailed and precise than those of any of its predecessors. By making clear what its obligations are, the treaty

reduces—it doesn't eliminate, but it reduces—the prospect that the Soviet Union will be tempted to exploit ambiguities or to take actions that may seem to fall in a nebulous "grey area."

The structure of the reductions will give us a basis to assess Soviet compliance early on. After the treaty has been in force for 30 days, we have the right to conduct onsite inspections to check the data the Soviets have provided us. From 90 days after entry into force, if the Soviets have any deployed shorter range missile outside a declared elimination facility, that will be a violation. The treaty requires the Soviet Union to destroy all its shorter range missiles within 18 months. From that time on, the existence of any Soviet shorter range missile will be a violation.

This "front-end loading" means that we will be able to see, from the outset, whether or not they are complying with treaty obligations of real military significance. The United States will be corralling and destroying shorter range systems in the same timeframe, but for us this will affect missiles in storage, not deployed systems.

The reductions in intermediate-range missiles will also be asymmetrical. Both sides will be reducing deployed intermediate-range systems from the outset, but we will keep a substantial force of Pershing IIs until well into the final months of reductions. This should provide a further incentive for the Soviet Union to comply with the timetable it agreed to. At no time during the period of elimination will we be left without missiles of our own.

Moreover, the verification provisions I outlined earlier are unprecedented. Increasing the likelihood of getting caught clearly decreases the attractiveness of cheating.

All of these safeguards are important. But no treaty in and of itself can fully guarantee compliance with its terms. No responsible U.S. official can tell you that we do not need to consider possible cheating.

Vigorous Reactions

We must react vigorously to questionable Soviet activities. We have to have the willpower to press them on our concerns. This Administration has not hesitated to do so. We have made energetic efforts to get the Soviets to take corrective actions. We have used diplomatic channels with the Soviets, we have reported the facts candidly to the Congress, and we have consulted extensively with our allies.

If we detect a Soviet action that seems in violation of their INF obligations, we will press them on it. Besides regular diplomatic channels in Washington and Moscow, we will now have a new compliance forum. Article XIII of the treaty establishes the Special Verification Commission or SVC. If we detect a suspicious activity, we will initiate an immediate meeting of the SVC. The Soviets cannot refuse.

At the SVC, we will have the right to describe our concerns and require a response. The Soviets may be able to demonstrate that the activity in question was not a violation. The SVC will be the place to determine that. If our concerns are not resolved, we will call for corrective action. If the Soviets were to persist in violating the treaty, we would have to react. There must be a penalty for cheating.

In this area, the Administration has shown its determination to take the actions necessary to safeguard our security. Faced with continuing Soviet refusal to live by the rules set by SALT II, the President took an appropriate and proportionate response. Specifically, he decided in May 1986 that future U.S. decisions on strategic forces would be based on the nature and magnitude of the Soviet threat—not on standards set by a treaty that was being selectively violated by the Soviet Union. He has implemented this decision with prudence and restraint.

We have identified the Krasnoyarsk radar as a clear violation of the ABM [antiballistic missile] Treaty. We will not agree to any further obligations regarding that treaty until that violation is dealt with satisfactorily.

Selective Violations

The pattern of Soviet behavior is not one of wholesale violations of arms control treaties. Their violations have been selective and specific. We must be prepared to respond specifically and selectively, with the focus squarely on what is in our own interests.

I believe that the INF Treaty is in the security interests of the United States. If the Senate does give the President its advice and consent to ratification, it will be because you agree with this judgment. And thus, if we ever were to determine that the Soviets were not complying with this agreement, this President, or any future President, should be able to count on the Senate's support to deal with that fact.

We as a country can safeguard our security only if we are willing to do what is needed to maintain and enhance our strength. If the Soviets cheat on this treaty, the President must be able to count on Congress to help him take the measures necessary to preserve our security and that of our allies.

For that is one of the key lessons of INF. If we have the willpower and the strength to see to our own security, we will succeed. If we do not, we have no one to blame but ourselves when we fail.

Lessons of INF

I have spoken at some length because I wanted to put the INF Treaty in the context of our overall approach to East-West relations. I wanted to explain how it fits into the strategy with which we and our allies have preserved peace in freedom for four decades, a strategy which the Soviets set out to challenge in the 1970s. I have made clear that this treaty is a substantial achievement. It strengthens Western military security. It sets new standards for arms control. It is a political success for the alliance. Our common achievement, based on close consultations, further strengthens the transatlantic partnership.

The INF Treaty, for all its benefits, leaves us with many challenges still unmet. As we and our allies move forward, we must draw the right lessons from INF. We succeeded in getting a good treaty because we persisted in our approach. If we and our allies had not gone forward with deployments, or if we had not had the courage of our convictions at the bargaining table, none of this could have been achieved.

There is much more to U.S.-Soviet relations than arms control, and we need to conduct those relations for the long term. We must not be misled by atmospherics. We need to watch the substance. It is just as dangerous to fall victim to euphoria when things are going well as it is to succumb to hysteria when there are problems. In dealing with the Soviet Union, INF shows that the key to success is strength, solidarity, and a clear-eyed understanding of what we ourselves want to accomplish. That is the only basis on which to build a structure of peace that will weather both good times and bad.

We and our allies face many challenges. NATO must continue to field the forces necessary to carry out its commitment to flexible response. There are substantial U.S. nuclear forces in place in Europe, including weapons on aircraft that can reach deep into the Soviet Union. Together with the U.S. troop presence, they constitute a convincing deterrent in their own right and a clear link to the U.S. strategic nuclear guarantee. In 1983, NATO adopted the Montebello nuclear modernization program, which they agreed was essential whether or not we reached an INF agreement. We must proceed with it, so that our nuclear forces in Europe maintain their capability to fulfill their mission of deterrence.

"The treaty shows the importance of negotiating from strength and of being patient, determined, and purposeful at the bargaining table."

This treaty does not resolve the conventional imbalance in Europe. That was not its objective. The security criterion which we established was that, in improving the nuclear balance, the treaty not harm NATO's conventional capabilities. The treaty more than meets that standard.

NATO has agreed that the next priorities in arms

control are 50% START cuts, a comprehensive global ban on chemical weapons, and establishing a stable and secure conventional balance in Europe by eliminating disparities. In all these areas, the INF Treaty sets precedents that will serve us well, including asymmetrical reductions and intrusive verification. And the treaty shows the importance of negotiating from strength and of being patient, determined, and purposeful at the bargaining table.

In START, as in INF, the initial concept proposed by the President is turning out to be the basis for a possible agreement. In the case of START, the President set the goal in his May 1982 proposal of significant reductions structured to enhance stability.

At each of his meetings with General Secretary Gorbachev, the President has moved us closer to this goal. At the December [1987] Washington summit, we agreed on essential constraints and counting rules and on verification guidelines that go even beyond the advances of INF. As we saw in INF, the indispensable next step is to translate these advances into treaty text. Doing so is a top priority for our team in Geneva.

Realism, Strength, and Dialogue

We have an ambitious agenda. The INF Treaty in no way signifies that we can relax our efforts. Instead, the treaty is a concrete demonstration of the fact that, in conducting our relations with the Soviet Union, we must hold true to our policy of realism, strength, and dialogue. And it also is a concrete demonstration that NATO, now nearly 40, with both the wisdom of time and the vigor of youth, continues to preserve peace in freedom for ourselves and for our allies.

George Shultz has been the US secretary of state since 1982.

"This treaty protects and preserves nuclear arsenals, instead of reducing them. This treaty is just a ruse by the Soviets."

The INF Treaty Threatens US Security

Jesse Helms

I want to talk about a highly-advertised, heavily promoted treaty that doesn't exist.

It doesn't exist, that is, in terms of what its proponents claim.

When the State Department negotiators went down to the wire negotiating the INF [intermediate-range nuclear forces] Treaty—and they were still negotiating it while they were walking up the steps to board the plane at Geneva to come home in December [1987]—their efforts were heralded as the greatest achievement since the Plan of Salvation.

Ever since, we have been bombarded with propaganda that this treaty "eliminated an entire class of nuclear weapons" and will bring about real reductions in nuclear warheads. The trouble is, it's not so.

We are constantly assured that this is "a breakthrough for arms control," which would carry us pell-mell into the arms of a START [strategic arms reductions talks] Treaty.

We are told that this treaty provides for "the most extensive system of verification ever proposed in an arms control treaty and the most intrusive plan for on-site inspection." That's not so either.

We are told that this treaty will strengthen NATO because, they say, it shows the collective unity of the Alliance and the willingness to make hard decisions—and back them up. Moreover, they claim, even after our brand-new Pershing II missiles are removed from Western Europe under this treaty, we will still have 4,000 nuclear warheads in Europe to stop Soviet aggression.

Party-Pooper

More than 50 Senators, and most other politicians, and practically all of the major liberal news media endorsed the treaty before they had even seen a copy of it.

But there's always one party-pooper in the crowd, and I guess I am it. I took the position that all Senators had a duty, under the advise and consent provision of the Constitution, to know what is in a proposed treaty before taking a position. I was willing to give the treaty a chance to (if you'll pardon the expression) show its stuff.

So I began asking questions—simple questions, based on fact, based on strange language in the treaty text, and based on numbers that simply don't add up. After all, when the waiter brings you the bill, it's always good to check the arithmetic.

The more I searched for the treaty that was being so highly advertised, the more impossible it became to find it. It was like the Cheshire Cat. It sat there grinning on the branch of the tree, but soon it faded away until only the smile remained.

Warheads Will Not Be Destroyed

The first question I asked, when I read the treaty text, was this: "What's this clause here that says that before the missiles are destroyed, the 'nuclear warhead device' and the electronic guidance equipment may be removed? Does that mean that no warheads are destroyed under this treaty? And if the warheads are not destroyed, how can it be correct to say that we are reducing nuclear arsenals for the first time?"

Well, you should have seen them when I asked those questions. I was likened to the skunk at a picnic. After all, hadn't the President said that the nuclear warheads would be destroyed? Hadn't the Vice President said that the nuclear warheads would be destroyed? Hadn't a dozen or more Senators danced around on the Senate floor and said that the nuclear warheads would be destroyed?

As it turned out, to put the most charitable face on, they were not very well informed. There it was—the no-warhead-destroyed loophole.

Jesse Helms, in a speech at Concord, New Hampshire, on February 9, 1988.

When I asked the Secretary of State [George Shultz], he looked embarrassed, and seemed to have trouble clearing his throat. Then he tried to explain. Of course, he said, the "fissionable materials" would be removed, but the vehicle that carries them would be "crushed."

I think I know an evasion when I see one, so I pressed the Secretary and the State Department negotiators harder. What are these "fissionable materials?" I asked. What the treaty says is "nuclear warhead device." And, isn't it possible just to take these nuclear warhead devices off the missiles to be destroyed and rebolt them on new missiles—different missiles not covered by this treaty, missiles targeted on Europe, or even the United States?

"In terms of explosive power, measured in kilotonnage, the Soviets have twelve times as much as the United States, and they would get to keep it all."

The diplomats were evasive on this point, but when it got down to the former Secretaries of Defense, and the Joint Chiefs of Staff—men familiar with nuclear weapons and how they are used, the answer was loud and clear. Yes, the warheads would *not* be destroyed, and they *could* be shifted to other missiles.

Oh, how the Soviet negotiators must have beamed when they got this loophole approved by the United States!

So what is going on here? In terms of numbers, we were told that the Soviets have deployed 3.4 times as many warheads that supposedly would be destroyed under this treaty as the United States. But that isn't so. They *won't* be destroyed. The Soviets will get to keep them all. In terms of explosive power, measured in kilotonnage, the Soviets have twelve times as much as the United States, and they would get to keep it all. Besides that, the nuclear warhead device and the electronic guidance system are obviously the most expensive part of the missile, the hardest to produce technologically, and the most limited in terms of production capacity. What a bargain for the Soviets!

We found out, therefore, that this treaty protects and preserves nuclear arsenals, instead of *reducing* them. This treaty is just a ruse by the Soviets to allow them to take out their old missiles, deployed since 1977, and rebolt those warheads to more modern Soviet missiles not covered by the treaty.

Hornswaggled in Geneva

On our side, our Pershing IIs are brand-new, state-of-the-art, and we would lose them if the treaty is ratified. We don't *have any other missile system under development* that could use those Pershing II warheads. We would have to take those warheads and reprocess the fissionable materials. If you accept the idea that the Pershing IIs should be destroyed, there is a slight advantage to us in reusing the fissionable material because we don't have enough capacity to make fissionable material in the first place.

So when you look at the bargain, it's a bad bargain. Our horse-traders got hornswaggled in Geneva, even if the Soviets keep their word, something they've never done!

In fact, if you go all the way back to 1920, you can find at least 30 major peace-keeping treaties that the Soviets have cheated on. The Soviets have always cheated, the Soviets will cheat in the future, and I have no doubt that the Soviets are cheating now. In fact, the Soviets cheat just to keep in practice.

Soviet Violations

With regard to arms control in particular, President Reagan has sent seven official reports to Congress giving the details of Soviet cheating. Let me give you a quick summary of the Soviet violations officially declared by the President:

1. SALT I ABM Treaty—now 10 confirmed violations;
2. SALT I Interim Agreement—5 confirmed violations;
3. SALT II Treaty—now 25 confirmed violations;
4. Limited Test Ban Treaty—over 30 confirmed violations;
5. Threshold Test Ban Treaty—over 24 probable violations;
6. Biological Warfare Convention—multiple confirmed violations;
7. Geneva Protocol on Chemical Weapons—multiple confirmed violations;

And last, but not least:

8. Kennedy-Khruschev Agreement—multiple confirmed violations. And I'll come back to that one.

It is important to remember that I am not talking about fringe or grey area violations. The President called them core area violations, fundamental to the integrity of the agreements. For example, the violations of the SALT II agreement include the illegal deployment of the SS-16, SS-24, the SS-25, and the SS-26. The SS-26 . . . is a 20-warhead weapon which is the most dangerous and most destructive weapon ever devised by man.

As [former] Secretary [of Defense Caspar] Weinberger testified before our [Senate Foreign Relations] Committee, "They have cheated in the past. I think we can expect they will cheat in the future whenever they decide it is in their national interest to do so."

All of this gets very interesting when we try to guess the likely areas where they will cheat. The main system the Soviets are supposed to eliminate is the SS-20. This is a missile which the Soviets began to deploy in 1977. Each missile has three warheads, and each of those warheads is seven times more powerful than the Hiroshima bomb. That's a lot of power, and its obvious purpose was to intimidate the Europeans into adopting a neutralist stance.

In fact, it turned out the opposite—after a tremendous fight in Europe to deploy NATO's own intermediate range missile, the Pershing II. *Our Pershing IIs don't have the firepower of the SS-20, and they only have one warhead.* But it is so fast, so accurate, and so certain of getting through Soviet missile defenses to reach targets deep in the Soviet Union that it has the Soviets scared to death.

Their main aim, of course, was to prevent the Pershing II from being deployed, and, having lost that fight, to get it out *now*. The INF Treaty does precisely that.

In return, we are supposed to get the elimination of the SS-20s. The problem is, the treaty does not really eliminate those SS-20s. Instead, we are allowing the Soviets to upgrade them to newer, and more effective missiles.

So, the question gets down to the perilous reality of this question: Are they really going to eliminate all their SS-20s? Or are they going to cheat, as Secretary Weinberger said?

So, as witnesses appeared to praise the treaty and themselves, I asked the obvious question. How many SS-20s are there that the Soviets are going to eliminate?

Well, I wish you could have seen them scurry for cover.

"The odds are that [the Soviets] . . . still have enough SS-20s to blow Europe to kingdom come—even if they do destroy all of the 650 missiles they have declared they have."

Senator, they said, the Soviets have *declared*, under the exchange of data, that they have 650 SS-20s.

"Yes, that is the number the Soviets *said*," I responded. But is that the total number they *really have*?

Senator, they said, we have "a high degree of confidence" in that estimate.

However, when I started to investigate, I found that a lot of responsible experts didn't have the same high degree of confidence in that number.

Admiral [William J.] Crowe, for example, in testimony before the Armed Services Committee,

said the Soviets could have as many as 300 more than they declared. Furthermore, the unclassified estimates of the Defense Intelligence Agency said the Soviets could have as many as 550 more than they declared. Other experts have reported the possibility that they might have produced as many as 1600 more. Admittedly, this number was based on some careful analysis of probabilities. Given the Soviet record of deceit and duplicity, the fact is that the Soviets could actually have two or three times as many more than they pretended to have in the official data exchange.

But whatever the actual number, the press has reported that the consensus of all our intelligence agencies—the four military agencies, and NSA [National Security Agency]—was unanimous in the National Intelligence Estimate in agreeing on the figure of 950, or 300 more than the Soviets declared. Only the CIA [Central Intelligence Agency] and the State Department disagreed. They tried to argue that the Soviets actually have 100 fewer SS-20s than they declared.

Which, to me, doesn't even make good nonsense.

Looking for Hidden Missiles

The sad truth is that we don't know, and we can't determine how many SS-20s the Soviets have. And the odds are that they still have enough SS-20s to blow Europe to kingdom come—even if they do destroy all of the 650 missiles they have declared they have.

Then, I thought, if they have this covert SS-20 force, surely we could detect it. So I asked where we could look for the SS-20s. It turns out, the only places we can look for SS-20s are the places where the Soviets tell us we can look. We can't look at any other place where we suspect they might be. Who is willing to assume that the Soviets will keep hidden missiles in places where the Soviets will allow us to look?

We don't know, and we can't know whether all the SS-20s have been destroyed. Even [Claiborne Pell,] the distinguished Chairman of the Senate Foreign Relations Committee—a strong supporter of the INF Treaty, said in the hearings: "My own view is that nobody knows and nobody will know." On this point, the Chairman and I are in full agreement.

The next question was a natural: If we do find them, what can we do about it? I was told that the treaty provides for "strong measures." We can protest. We can write letters. We can make strong representations in the monitoring commission. We might even find a way to flog the Soviets with a wet noodle.

Sweeping Across Europe

Meanwhile, the Soviet armies could very well be sweeping across Europe even before our diplomats get their bags packed to go to Geneva for more negotiations.

The Soviets understand United States policies very well. They've had years of practice in orchestrating political actions here. They know how to manipulate the media. They know how to influence public opinion. The Gorbachev road show proved that. He left America in a state of euphoria bordering on adulation of this "nice, friendly fellow from Moscow."

The Soviets know one further thing: With the U.S. Congress as presently constituted, the Soviets are supremely confident that there will not be a strategic nuclear response by the United States when the Soviets move across the borders to take over the NATO countries with the conventional military superiority the Soviets already possess.

So the Soviets know that once this treaty is fully implemented completely, there will be no real deterrent. What is a "real deterrent"? It is one that keeps the Soviets from even thinking about war—not one that might make it rough on them after they have launched an attack.

"Cheating on the INF Treaty means illegal nuclear weapons within range of Miami, Charleston, Washington, Baltimore, Philadelphia, New York, Boston, even Concord, New Hampshire."

Once the Pershing IIs are gone, we can look forward to startling changes in Europe—a unified and neutralized Germany beholden to the Soviet Union, one-party governments in other key European states, the economic integration of the European Economic Community with the Soviet/Warsaw pact economic group called COMECON. This would be a world in which the very idea of NATO would be irrelevant.

So what happened to that wonderful treaty we've heard so much about?

I keep looking, hoping there will be some saving provision that makes it worthwhile, instead of deadly dangerous. But every time I do that, whole new problems arise.

For example, I tried to check a few things with Admiral Crowe, the Chairman of the Joint Chiefs of Staff.

Admiral, I asked, this treaty prohibits INF missiles all over the world, including the SS-4 and the SS-5, doesn't it? Oh, yes, he reassured me; that's one of its great advantages.

And we can inspect any base where those missiles have been detected in the past, can't we? I asked.

He hesitated, and I began to get worried again. Then he answered: Only if they are specified in the treaty.

Well, I said, in October 1962, President Kennedy declared in a nationwide television address that the Soviets had installed SS-4s and SS-5s in Cuba. Did we ever get on-site inspection of those bases?

The Admiral seemed to be stumped. But the fact was that the Soviet Union promised on-site inspection (under UN supervision), and then reneged on the deal.

Why aren't those Cuban bases in the INF treaty? Didn't the Kennedy-Khrushchev Agreement say that the Soviets couldn't reintroduce nuclear weapons into Cuba? Didn't we know that intermediate range nuclear weapons of one type or another had been introduced into Cuba in violation of the Kennedy-Khrushchev Agreement? Hadn't President Reagan twice said that the Agreement had been violated? Hadn't the Director of the CIA, the Under Secretary of Defense, and the President's General Advisory Committee on Arms Control all said that the Soviets were in violation of the Kennedy-Khruschev Agreement? Did not the Soviets introduce a missile in Cuba that was an almost identical version to one supposedly forbidden in the INF Treaty? Couldn't the Soviets fly in SS-20s and their launchers in the cargo bay of their giant Condor air transports, off load them at night, and we would have no way to detect the violation, no mechanism under the treaty to inspect or correct it? To all these questions, the Admiral gave a grudging yes.

Almost by chance, I had stumbled on the Cuba loophole. Apparently no one in the Administration had ever thought of it before. It is one thing to talk about the dangers of cheating on an INF Treaty in Europe, which is far away to many Americans. But it is another thing when cheating on the INF Treaty means illegal nuclear weapons within range of Miami, Charleston, Washington, Baltimore, Philadelphia, New York, Boston, even Concord, New Hampshire.

So that is where we stand. I am still looking for that treaty we were promised. I still can't find it. If they can show it to me, I might be persuaded to vote for it. But somehow, I just don't know what happened to it at all.

Fighting for Freedom

If the American people could be properly informed and sufficiently motivated about what this treaty is, and what it is not, they could demand that the treaty be rewritten and renegotiated to protect, not endanger, the cause of freedom in the world. . . .

The question is: Do we care enough to keep it?

Jesse Helms is a Republican senator from North Carolina.

"After INF treaty ratification ... inspectors will visit every site listed in the treaty's 100-page 'memorandum of understanding' or appendix to verify its accuracy."

INF Treaty Compliance Can Be Verified

R. Jeffrey Smith

The impact of the U.S.-Soviet treaty eliminating intermediate-range nuclear forces (INF), signed in Washington by President Reagan and Soviet leader Mikhail Gorbachev, will probably be felt first at a military base in the city of Kapustin Yar, where the Soviet Union launched its first ballistic missile 40 years ago.

One of the INF treaty's most unorthodox provisions will unfold at the base, 660 miles southeast of Moscow, shortly after the pact takes effect, when the Soviets begin launching unarmed SS20 missiles eastward virtually around-the-clock just to get rid of them.

Similar launches of powerful, highly accurate Pershing II missiles, without warheads, from Cape Canaveral east over the Atlantic may be made by the United States if studies show this is the cheapest, safest way to destroy the $6 million rockets within the treaty's three-year deadline.

The 2,800 U.S. and Soviet missiles to be destroyed under the INF treaty are only a fraction of those in the countries' nuclear arsenals, and their destruction will have only modest impact on the overall nuclear balance. But no previous arms agreement has called for destruction of so many weapons in such a brief period. Because of this and because of the novelty of its on-site verification procedures, the INF treaty is a departure for the superpowers.

Every U.S. and Soviet land-based missile designed to fly between 300 and 3,400 miles will be eliminated under the treaty, including Soviet SS4, SS5, SS12, SS20 and SS23 missiles, and U.S. Pershing IA, Pershing II and ground-launched cruise missiles, most of which were produced and deployed within the last decade.

None of the Soviet arms is aimed at the United States. All are pointed at U.S. allies and China. However, all but a few of the U.S. arms governed by the pact are aimed at the Soviet Union.

A Novel Approach

To swiftly wipe out whole categories of medium-range and shorter-range missiles under the treaty's elaborate requirements, both sides will take novel steps and encounter odd headaches.

For example, they will face environmental challenges in burning tons of highly toxic rocket propellant. They will have to arrange for disassembly of perhaps 2,000 modern nuclear warheads and missile guidance mechanisms, and return the radioactive materials to weapons plants or nuclear reactors. And they will have to build special housing for military inspectors to be stationed outside highly sensitive military facilities in both countries for 13 years.

It took six years of halting, fitful superpower negotiations marked by angry denunciations and florid rhetoric in Moscow and Washington to produce the treaty, 41 pages of legalistic text. Two "protocols," or supplements to the treaty, totaling another 78 pages, spell out in extraordinary detail how the INF weapons must be dismantled and how each side can ensure the other has met its obligations. These procedures call for the most intrusive inspections of any arms pact since the dawn of the nuclear age, costing both governments the equivalent of hundreds of thousands of dollars annually.

"The inspection protocol makes the tax code look simple," one official says. For example, when the Soviets slice the back end off certain trucks to make them incapable of carrying SS20 missiles, U.S. inspectors will be watching to ensure that the pieces are 39 inches long, not 38. When Soviet SS12 and SS20 missiles are launched from Kapustin Yar, U.S. officials will be there to inspect them beforehand

R. Jeffrey Smith, "A Treaty That Beats Weapons Into Waste," *The Washington Post National Weekly Edition*, December 21, 1987. © The Washington Post.

and observe them arching high above the Kazakh plain.

In an unprecedented, 114-page burst of candor, the two sides have also provided spare descriptions of every site where INF weapons are produced, assembled, maintained, stored, maneuvered for training and deployed for war. Included are sites in at least eight states, as well as eight European countries and at least four Soviet republics, newly subject to intrusive inspections.

Together with treaty provisions allowing dozens of inspections annually, the detailed disclosure of previously sensitive military data is widely considered one of the treaty's most significant accomplishments.

The disclosure runs counter to a decades-old U.S. policy of refusing to acknowledge or disclose the presence of nuclear weapons anywhere in the world. It also breached Soviet policies that one negotiator says called for anyone who supplied such data to be shot for "committing treason."

"Outside the main rail entrance to the [Soviet SS20 missile factory in Votkinsk] . . . Americans will weigh, measure and perhaps x-ray anything large enough to look like an SS20."

This legacy of military secrecy had to be overcome, both sides realized, in order to establish on-site inspection procedures that would eliminate the risk of militarily significant cheating. Concerns about cheating were particularly high because all but a few INF missiles can be trucked from one site to another to avoid detection.

Previous nuclear arms agreements have constrained large, intercontinental ballistic missiles deployed in concrete-reinforced silos, readily identifiable in reconnaissance satellite photos. But the INF weapons are so compact that—until now—neither side knew exactly how many the other had or where they were deployed at any given moment.

The Reagan administration has repeatedly said, for example, that the Soviet Union had deployed 130 to 140 launchers for shorter-range missiles. U.S. officials say newly furnished Soviet information lists 197 launchers and 387 deployed missiles. A secret appendix to the treaty also shows that the United States has deployed dozens more missiles in Europe than it has previously acknowledged, according to U.S. officials.

The decision to eliminate all INF weapons, those warehoused as well as those deployed at readily observable missile bases, is a calculated risk. The

provision blocks either side from hiding hundreds of missile-carrying trucks in warehouses and garages for sudden use in war.

But neither side can offer complete assurance that stored missiles have been eliminated, simply because hiding them is too easy and finding them too hard.

"Nothing is 100 percent perfect," Secretary of State George P. Shultz said. "It's possible" for the Soviets to cheat under the treaty provisions, "but I think, if it occurs, it would be in very small proportion."

Chief U.S. arms negotiator Max M. Kampelman says, "You have to ask yourself—will they cheat? And our conclusion was [that whenever] they can cheat—and you assume they will—it won't hurt us. And so we say under those circumstances, we recommend this treaty."

These statements signal an important shift in the administration's position of verification that is symbolized by the INF treaty.

Instead of airtight procedures for verifying Soviet compliance, a goal conservatives have long espoused, the treaty's procedures make cheating difficult, but not impossible. Instead of ruling out any risk of the slightest Soviet violation, the treaty aims to block only militarily significant cheating.

"This treaty is ultimately based on the same verification standards applied by Reagan's predecessors, both Democrat and Republican," says James P. Rubin, assistant director of the private Arms Control Association in Washington. "All the administration's talk about perfect verification turns out to have been empty rhetoric, and rightly so."

Inspecting Every Site

One to three months after INF treaty ratification by the U.S. Senate and the Supreme Soviet, teams of inspectors will visit every site listed in the treaty's 100-page "memorandum of understanding" or apendix to verify its accuracy.

U.S. officials say both sides viewed these preliminary inspections apprehensively, causing each to delay the exchange of certain data to mutual benefit.

For example, the Defense Department moved items covered by the treaty away from one highly sensitive facility, so that it would not be listed and inspected by the Soviets. U.S. intelligence officals say the Soviets dismantled some INF facilities and moved missiles to new locations to evade U.S. inspections.

"These actions were militarily sound and completely legal," even if they were not in keeping with the pact's new spirit of openness, one official says.

With initial inspections completed, each country will officially declare where its missiles are to be eliminated. The likely U.S. sites are Army bases in Pueblo, Colo., and Tooele, Utah, where the rockets would be burned in specially designed pits. Alternatively, the two sides could strap the missiles down and simply ignite the propellant, providing a fiery symbol of arms under control. But specialists say these techniques may create too much air pollution, causing both sides to consider destroying the rockets by launching from existing test ranges.

The treaty allows up to 100 rockets to be destroyed this way, falling harmlessly back to Earth. But all launches must be completed within six months after the treaty takes effect, and no more than four missiles can be launched each day. The United States insisted on these constraints to block the Soviets from gaining any military advantage from the launches.

No other flight tests are allowed, a constraint that can be readily observed by U.S. satellites and ground radars on the perimeter of the Soviet Union. The aim of this provision, according to Shultz and other officials, is to limit the military value of secretly produced missiles. "If there are no tests, before long the system becomes obsolete," Shultz says.

Secret production of the Soviet SS20 missile at an existing factory in Votkinsk, near the Ural Mountains, is to be blocked by stationing U.S. inspectors outside the main rail entrance to the plant for 13 years. Somewhat like interstate highway inspectors, the Americans will weigh, measure and perhaps x-ray anything large enough to look like an SS20.

In so doing, they will pick up fresh information about another, larger Soviet missile produced at the plant, the SS25, which is technically not covered by the agreement. The Soviets allowed such inspections when the United States agreed to permit the Soviet Union to station a weighing and measuring team outside a former Pershing II production facility in Magna, Utah, which now produces MX and Trident II missiles.

Scheduled Elimination

U.S. negotiators persuaded the Soviets to eliminate all shorter-range missiles within 18 months, and eliminate all but 180 deployed medium-range missiles within 2 1/2 years. Otherwise, each side pursues its own schedule, waiting either until the last minute or racing to finish first.

Besides eliminating missiles, the two sides are also required to dismantle and destroy all related missile launching equipment within three years. Initially, the two sides planned to destroy hundreds of mobile missile-carrying trucks, but the Army decided it wanted to keep the tractors used to pull the Pershing IIs through German forests, and the Soviets decided they wanted to give their SS20 flatbeds to civilians. The two sides then entered protracted negotiations over how much of the flatbeds must be lopped off to make them too short and too weak to carry an SS20, finally settling on one meter.

At the end of the three-year period for eliminating missiles, launchers and support facilities, each side

has the right to conduct an additional inspection of the missile bases, factories, storage sites and training areas to ensure that the other has met its obligations.

In the meantime, each country is allowed to send inspection teams to 20 missile bases and related facilities annually to ensure that dismantling is proceeding according to promise. An argument is under way within the administration whether U.S. teams should be run by the military or the Arms Control and Disarmament Agency. If the inspection is conducted on Soviet territory by the United States, for example, the team will first be sent to Moscow or Irkutsk with measuring devices, radiation detectors and special dual-lens cameras that instantly produce two identical prints—one for the Soviets and one for the Americans. At least two of the 10 inspection team members must speak Russian.

The Soviets are permitted to inspect the equipment to ensure it has no hidden espionage capabilities. They must then arrange transportation within nine hours to any site the team identifies, weather permitting.

Inspection

Once at the site, the Soviets take the pictures, but must photograph whatever the U.S. side demands, an unusual arrangement that reflects lingering uneasiness on both sides about providing unrestricted access. Soviet inspections on U.S. territory are governed by identical rules.

"Each country is allowed to send inspection teams to 20 missile bases and related facilities annually to ensure that dismantling is proceeding according to promise."

The Soviets agreed to a U.S. demand for 15 such inspections annually for the first five years after all INF weapons are eliminated, and 10 inspections annually for the next five years. U.S. officials believe that any Soviet missiles remaining after 10 years will be obsolete.

Two brief, legally binding notices in the treaty documents will provide U.S. and Soviet assurance that military inspections on the territory of their allies will comply with appropriate national laws. No more than half of the inspections can be conducted in any single country each year.

R. Jeffrey Smith is a staff writer for The Washington Post National Weekly Review.

INF Treaty Compliance Cannot Be Verified

National Security Record

On December 8, 1987, Ronald Reagan and Mikhail Gorbachev signed in Washington the Treaty for the Elimination of Intermediate-Range and Shorter-Range Missiles, known as the INF agreement. It includes unprecedented on-site inspections inside the Soviet Union and will eliminate a whole category of nuclear missiles. Yet a formidable array of critics have expressed serious reservations. The treaty, they contend, is flawed and would leave our NATO allies at risk.

What is this treaty that has generated such controversy? For one thing, it is a long, complex and ambiguous legal document. The treaty is in four parts: the 19-page treaty itself; a 16-page protocol on the procedures governing the elimination of missiles; a 21-page protocol outlining the inspection procedures; and a 73-page memorandum of understanding that lists the numbers and locations of the missiles that are to be eliminated.

The main points of the agreement are:

• It eliminates three entire categories of weapons worldwide, although most are in or facing Europe. They are:

—Intermediate-range ballistic missiles (LRINF weapons) with ranges from 600 to 3,400 miles.

—Shorter-range ballistic missiles (SRINF weapons) with ranges from 300 to 600 miles.

—Ground-launched cruise missiles (GLCMs) with the same range capabilities.

• The operational weapons to be eliminated carry 1,797 warheads on the Soviet side and 429 on the U.S. side. However, none of the warheads actually will be eliminated, only the missiles that carry them.

• The LRINF missiles are to be destroyed within three years. The shorter-range missiles must be removed from operational status within 90 days and

destroyed within 18 months. Destruction is to be by explosion, burning or cutting into pieces, but 100 may be destroyed by launching.

• Neither side can produce or possess the classes of weapons eliminated for the duration of the treaty, which is indefinite.

• Neither side may produce or flight test any of the banned missiles, or produce their launchers.

• Each side may establish teams of inspectors to monitor verification, including the stationing of inspectors at a missile facility on the other's territory to verify that the treaty is not being violated. The total inspection force on each side may not exceed 600; 200 on-site in the other's country, 200 engaged in challenge inspections of agreed sites and 200 air crew and other transportation personnel.

The weapons to be eliminated are:

Soviet Weapons	Range	No. Operational & Spares	In Storage	Total
SS-20	3,100 mi.	470	245	715
SS-4	1,600 mi.	0	105	105
SS-5	2,100 mi.	0	6	6
SS-12	585 mi.	220	506	726
SS-23	325 mi.	167	33	200
SSC-X-4	1,800 mi.	0	84	84
Total USSR				1,836
U.S. Weapons				
Pershing II	1,200 mi.	120	260	380
GLCM	1,550 mi.	309	0	309
Pershing IA	467 mi.	0	170	170
Pershing IB	600 mi.	0	0	0
Total U.S.				859

Numbers from the Soviets

All of the weapons to be banned are ballistic missiles except for two ground-launched cruise missiles: the U.S. GLCM and the Soviet SSC-X-4, a long-range GLCM that is still experimental. The numbers in the chart above are from the memorandum of understanding that is part of the

National Security Record, "The INF Treaty," January 1988. Reprinted with permission.

treaty. However, the totals listed in the treaty do not include 84 Soviet SSC-X-4 cruise missiles at the missile storage facility at Jelgava in the USSR. The existence of these 84 long-range cruise missiles was unknown until the Soviets provided the data, to the surprise of the U.S. intelligence community. Although not included in the official numbers, the SSC-X-4s are subject to destruction under the treaty.

The numbers and locations of Soviet missiles were provided by the Soviets. Those for the SS-20 are close to the CIA [Central Intelligence Agency] estimate but less than other estimates. As a result, there is some speculation that Moscow simply confirmed the lower CIA estimate. The actual number, especially of spares, remains in doubt.

"The U.S. is taking Moscow's word for the numbers and locations of spare missiles, and there are no provisions to inspect 'suspect' Soviet facilities that are not on the agreed list of places that can be inspected."

All of the weapons to be eliminated carry a single nuclear warhead except the Soviet SS-20, which carries three. The SS-4, SS-5 and Pershing IA are aging weapons that have been retired from operational service by the U.S. and USSR. However, a number of SS-4s and Pershing IAs are kept in storage. They also will be destroyed under the treaty.

In addition to the U.S. and Soviet weapons to be destroyed, the West German government has agreed to eliminate 72 Pershing IAs that have long been deployed in Europe. The U.S. maintains control of the nuclear warheads for these German Pershing IAs. The Soviets insisted on their elimination under the INF agreement, but the U.S. rejected this Soviet demand on the grounds that the German government is not a party to the agreement. The U.S. then quietly pressured Bonn to make a unilateral concession on these weapons, and Chancellor Helmut Kohl eventually agreed to eliminate the West German Pershing IAs as soon as the U.S. and Soviet INF weapons have been eliminated. The U.S. will then return the warheads to the United States.

The Pershing IB does not exist. It has been proposed as a shorter-range modification of the Pershing II, but this option is now precluded by the agreement.

The protocol that sets forth the procedures to be followed in the elimination of missiles provides for the withdrawal of complete units to simplify verification. Destruction is limited to agreed locations. Since all LRINF and SRINF weapons are to be eliminated, the verification of compliance is easier than if some were to be retained.

The most controversial part of the INF agreement is the section on verification, since there are serious doubts that the numbers and locations of small mobile weapons can be verified with any meaningful degree of accuracy. The administration, therefore, has devised very detailed verification procedures that are the most intrusive ever agreed upon.

The principal method of monitoring compliance will continue to be the use of National Technical Means (NTMs), which include satellite photography, electronic and communications intercepts, data collection by aircraft and other methods that involve "technical" rather than "human" collection of information. But the agreement also includes new and innovative on-site inspection provisions, principally:

Baseline Inspections. On-site inspections of agreed locations may be conducted, including bases in East Germany and Czechoslovakia, in addition to the Soviet Union. In-country transportation for inspectors will be provided by the host country.

Close-out Inspections. As missile facilities are eliminated, they can be inspected to verify compliance.

Elimination Inspections. Each side may observe the destruction of missiles and launchers at agreed-upon sites.

Short-notice or Challenge Inspections. For 13 years after the treaty is ratified and enters into force, each side may conduct a specified number of on-site inspections of agreed locations each year. The inspectors must give 16 hours notice before arriving at the city of entry and 6 hours notice before departing for the site to be inspected.

Portal Monitoring of Production. The U.S. will have a continuous monitoring system and team of inspectors at the final assembly plant for the Soviet SS-25 ICBM at Votkinsk in the Ukraine. The banned SS-20 INF missile and the allowed SS-25 ICBM are externally very similar, and Votkinsk previously was a final assembly facility for the SS-20. The inspectors will try to assure that the Soviets are not building SS-20s instead of SS-25s at Votkinsk. Soviet on-site monitoring will be conducted at the Hercules plant at Magna, Utah, once a Pershing II assembly plant but now making components of the MX missile.

Special Verification Commission. A special commission is to be established to try to resolve compliance issues.

The SS-25 Problem

Despite the "strict" verification, there are a number of problems. House Armed Services Committee Chairman Rep. Les Aspin (D-Wis) shares the concern of others that Moscow could "cheat legally" by simply retargeting longer-range missiles on the INF targets in NATO. Aspin also has focused on the two glaring weaknesses in the elaborate verification system: the U.S. is taking Moscow's word for the numbers and locations of spare missiles, and there are no provisions to inspect "suspect" Soviet facilities that are not on the agreed list of places that can be inspected.

The similarity of the SS-20 and the SS-25 ICBM is a major problem. The first stage of the two missiles is identical. The second stage, the Soviets say, has different dimensions, but no American has ever seen, let alone measured, either missile. The SS-25 launcher has 7 axles instead of 6 on the SS-20 launcher. The canisters that carry the missiles are similar. SS-20s could be carried in SS-25 canisters on SS-25 launchers and no one would know.

In the negotiations, the Soviets refused to agree to on-site inspections of SS-25 bases. Under pressure of the deadline, the best the U.S. negotiators could get on this issue was a Soviet agreement to open the roofs of the SS-25 garages on 6 hours notice and leave the launchers in the open for 12 hours. But what does that tell? Only that the Soviets have put a certain number of SS-25s in the open. It does not tell whether SS-20s are in SS-25 canisters, nor how many other SS-20s or SS-25s there may be underground or hidden in other locations. The compromise U.S. negotiators made on this issue under White House pressure to get an agreement by December 7 [1987] appears to be worthless.

Other verification weaknesses are that only Soviet transportation can be used by inspectors inside the Soviet Union, and if U.S. inspectors want pictures they can only be taken by Soviets using Soviet cameras. This puts U.S. inspectors totally at the mercy of their Soviet "guides."

Cruise Missiles

The possibility of banning non-nuclear GLCMs was a major issue between [the departments of] State and Defense [in 1987]. The U.S. has a considerable technological advantage in GLCMs, one of the best options for strengthening NATO's conventional deterrence. They offer the possibility of delivering a non-nuclear explosion with great accuracy on the air bases, bridges, choke points and other priority targets the West would have to hit in the event of a Soviet non-nuclear assault. But they present a problem for arms control: they can carry either a nuclear or conventional warhead and the warheads can quickly be changed from non-nuclear to nuclear.

When this issue went to the president [in] July [1987], he chose the arms control option—State over Defense—and agreed to ban all long-range GLCMs. This decision prevents the West from deploying a weapon in which it has a significant technological advantage. It also presents opportunities for Soviet cheating: despite the ban, GLCMs could be tested as air-launched or sea-launched cruise missiles.

The treaty bans missiles and cruise missiles with ranges from 300 to 3,400 miles. But it is not possible to determine accurately the full range of these weapons. For treaty purposes, the range of a weapon is the maximum range to which it has been tested. The treaty limits can be undermined by extending or reducing the ranges of permitted missiles.

A basic problem is the lack of hard information on Soviet production. Little is known about the rate of Soviet missile production, the number of missiles in storage, the use of floor space in Soviet production plants, or even the existence of some production facilities.

Finally, the compliance mechanism chosen for this agreement is a new Special Verification Committee. While not defined, it sounds suspiciously like the Standing Consultative Commission of the SALT and ABM agreements, which has been unable to correct Soviet violations but which has kept its deliberations—and its failures—highly classified and away from public scrutiny.

A Shell Game

The INF treaty was concluded in the worst possible way: it was negotiated under pressure of a deadline. Potentially serious concessions appear to have been made. Frank Gaffney, the acting assistant secretary of defense who fought for strict verification only to be forced from office after the departure of Defense Secretary Caspar Weinberger, said the provisions and negotiating history of the highly controversial ABM treaty are models of clarity compared to the ambiguities of the INF treaty. The lack of on-site inspection of suspect sites means, according to Gaffney, that the U.S. will be playing a shell game on verification, allowed only to look under those shells the Soviets say can be inspected.

"The INF treaty was concluded in the worst possible way: it was negotiated under pressure of a deadline."

In its ratification deliberations the Senate must consider these issues, together with the effect on NATO and the overall strategic balance, before deciding whether to commit America's future security to this agreement.

National Security Record *is a monthly newsletter that reports on Congress and national security affairs. It is published by the Heritage Foundation, a conservative think tank in Washington, DC.*

bibliography

The following bibliography of books, periodicals,
and pamphlets is divided into chapter topics
for the reader's convenience.

Arms Control

Gordon Adams
''What Next for Arms Control?'' *Dissent*, Spring 1988.

Leon Aron
''What Gorbachev Is Saying About the US,'' *Backgrounder*, November 5, 1987. Available from The Heritage Foundation, 214 Massachusetts Ave. NE, Washington, DC 20002.

Henrik Bering-Jensen
''A Dissent over Arms Control,'' *Insight*, December 28, 1987/January 4, 1988.

Bruce D. Berkowitz
Calculated Risks. New York: Simon & Schuster, 1987.

Hans Bethe
''Containing Deterrence,'' *New Perspectives Quarterly*, Spring 1987.

Tom Bethell
''Arms-Control Frenzy,'' *National Review*, October 23, 1987.

Mary Tedeschi Eberstadt
''Arms Control and Its Casualties,'' *Commentary*, April 1988.

Colin S. Gray
''The Gorbachev Offensive,'' *Society*, July/August 1987.

Morton H. Halperin and Madalene O'Donnell
''The Nuclear Fallacy,'' *Bulletin of the Atomic Scientists*, January/February 1988.

Michael Johns
''Peace in Our Time,'' *Policy Review*, Summer 1987.

Victor Karpov, interviewed by Konstantin Isakov
''An Equation Without Unknown Quantities,'' *New Times*, no. 10, March 1988.

Henry A. Kissinger
''A New Era for NATO,'' *Newsweek*, October 12, 1987.

Henry A. Kissinger
''The Singular Threat to the Atlantic Alliance,'' *National Review*, July 3, 1987.

Morton M. Kondracke
''The Reagan Method,'' *The New Republic*, November 30, 1987.

Georg Kwiatowski and Unni Krishnan
''Disarmament Is Possible,'' *World Marxist Review*, February 1988.

Michael MccGwire
Military Objectives in Soviet Foreign Policy. Washington, DC: The Brookings Institution, 1987.

Robert S. McNamara
''Blundering into Disaster: The First Century of the Nuclear Age,'' *The Brookings Review*, Spring 1987.

The New York Times Magazine
''Symposium: A World Without Nuclear Weapons?'' April 5, 1987.

Sam Nunn
''Arms Control in the Last Year of the Reagan Administration,'' *Arms Control Today*, March 1988. Available from The Arms Control Association, 11 Dupont Circle NW, Washington, DC 20036.

Ronald E. Powaski
''Is Nuclear Deterrence Immoral?'' *America*, May 16, 1987.

Eugene V. Rostow
''There Is No Alternative Strategy,'' *Global Affairs*, Spring 1988.

Edward L. Rowny
''Hard Word Ahead in Arms Control, *Department of State Bulletin*, January 1988.

Theo Sommer
''Why Fear Negotiations on Conventional Forces?'' *World Press Review*, May 1988.

Roland M. Timerbaev
''A Soviet Official on Verification,'' *Bulletin of the Atomic Scientists*, January/February 1987.

Stansfield Turner
''Winnowing Our Warheads,'' *The New York Times Magazine*, March 27, 1988.

Economics of the Arms Race

Gordon Adams
''Economic Conversion Misses the Point,'' *Bulletin of the Atomic Scientists*, February 1986.

Richard Brookhiser
''Rescuing the Military,'' *National Review*, February 14, 1986.

Michael Closson
''A Disarming Solution,'' *Multinational Monitor*, February 1988. Available from P.O. Box 19405, Washington, DC 20036.

Robert B. Costello
''Acquisition Reform Moves . . . Front & Center,'' *Defense 88*, January/February 1988. Available from the Superintendent of Documents, U.S. Government Printing Office, Washington, DC 20402.

James Fallows
''The Spend-Up,'' *The Atlantic Monthly*, July 1986.

Philip Gold
''Defense Reform Is Assured, but in What Sort of Form?'' *Insight*, March 9, 1987.

Richard Halloran
To Arm a Nation. New York: Macmillan Publishing Company, 1986.

Nick Kotz
''The Chesapeake Bay Goose Hunt, the Beautiful Secretary, and Other Ways the Defense Lobby Got the B-1,'' *The Washington Monthly*, February 1988.

Michael Mecham	"Panel Finds Defense Policies Erode U.S. Technology Base," *Aviation Week & Space Technology*, May 18, 1987.	Richard Perle	"The INF Treaty: Peace in Our Time?" *American Legion Magazine*, April 1988.
Rosy Nimroody	"The S.D.I. Drain," *The Nation*, January 16, 1988.	Richard Perle	"What's Wrong with the INF Treaty," *U.S. News & World Report*, March 21, 1988.
Stan C. Pace	"The Military Budget Helps the Economy," *The New York Times*, March 6, 1988.	Ronald E. Powaski	"An Important Step Toward Nuclear Sanity," *America*, October 3, 1987.
John Ralston Saul	"The War Business," *World Press Review*, June 1987.	Michael G. Renner	"Disarming Implications of the INF Treaty," *World Watch*, March/April 1988.
Robert J. Shapiro	"A Frightening New Numbers Game," *U.S. News & World Report*, September 28, 1987.	John Silber	"The Illusion of Peace: The INF Treaty," *Vital Speeches of the Day*, February 15, 1988.
Lee Smith	"How the Pentagon Can Live on Less," *Fortune*, July 21, 1986.	E.P. Thompson	"The Peace Movement's Next Task," *The Nation*, December 12, 1987.
Harlan K. Ullman	"National Security and Fiscal Reality: An Impending Collision," *The World & I*, May 1988.	Philip Towle	"How Good Is the INF Treaty?" *World Press Review*, April 1988.
Caspar Weinberger	"Technological Leadership, the Industrial Base & National Security," *Defense 87*, July/August 1987.	Henry Trewhitt	"Arms Control: Is It Good for Us?" *U.S. News & World Report*, December 14, 1987.
Jerome B. Wiesner	"More R & D in the Right Places," *Issues in Science and Technology*, Fall 1987.	Paul C. Warnke	"INF Treaty a Good Start," *Bulletin of the Atomic Scientists*, March 1988.
		Russell Watson	"At Long Last, an Arms Deal," *Newsweek*, September 28, 1987.

The INF Treaty

Kenneth L. Adelman	"Why an INF Agreement Makes Sense," *Vital Speeches of the Day*, June 15, 1987.
Les Aspin	"Unilateral Moves for Stability," *Bulletin of the Atomic Scientists*, December 1987.
Bruce D. Berkowitz	"An INF Treaty Discredits Arms Control and Promotes Conflict," *Orbis*, Winter 1988.
Tom Bethell	"Arms Control: The Untold Story," *National Review*, February 19, 1988.
Vyacheslav Boikov	"Looking for a 'Compensation,'" *New Times*, no. 5, January 1988.
Bulletin of the Atomic Scientists	"Six minutes to Midnight," January/February 1988.
Frank C. Carlucci	"How the Intermediate-Range Nuclear Forces Treaty Supports US Strategy," *Defense*, March/April 1988.
Congressional Digest	"The Intermediate-Range and Shorter-Range Missiles Treaty: Pro & Con," April 1988.
Lynn E. Davis	"Lessons of the INF Treaty," *Foreign Affairs*, Spring 1988.
Yuri V. Dubinin	"A Nuclear-Free and Safe World: The Soviet Concept," *Soviet Life*, February 1988.
Frank J. Gaffney Jr.	"A Layman's Guide To Fixing the INF Treaty," *National Review*, March 18, 1988.
Frank J. Gaffney Jr.	"A Policy Abandoned," *The National Interest*, Spring 1988.
Jack Kemp	"Arms Control Perverted," *National Review*, May 22, 1987.
Melvyn Krauss	"Arm Wrestling over Europe," *Reason*, April 1988.
Greg LeRoy	"After the INF Treaty," *Science for the People*, March/April 1988.
Carnes Lord	"The INF Hard-Sell," *The American Spectator*, March 1988.

index